THE REALITY FRAME

RELATIVITY AND OUR PLACE IN THE UNIVERSE

BRIAN CLEGG

ICON

This edition published in the UK in 2017
by Icon Books Ltd, Omnibus Business Centre,
39–41 North Road, London N7 9DP
email: info@iconbooks.com
www.iconbooks.com

First published in the UK in 2017 by Icon Books Ltd

Sold in the UK, Europe and Asia
by Faber & Faber Ltd, Bloomsbury House,
74–77 Great Russell Street,
London WC1B 3DA or their agents

Distributed in the UK, Europe and Asia
by Grantham Book Services,
Trent Road, Grantham NG31 7XQ

Distributed in the USA
by Publishers Group West,
1700 Fourth Street, Berkeley, CA 94710

Distributed in Canada by Publishers Group Canada,
76 Stafford Street, Unit 300,
Toronto, Ontario M6J 2S1

Distributed in Australia and New Zealand
by Allen & Unwin Pty Ltd,
PO Box 8500, 83 Alexander Street,
Crows Nest, NSW 2065

Distributed in South Africa
by Jonathan Ball, Office B4, The District,
41 Sir Lowry Road, Woodstock 7925

Distributed in India by Penguin Books India,
7th Floor, Infinity Tower – C, DLF Cyber City,
Gurgaon 122002, Haryana

ISBN: 978-178578-281-7

Typeset in Bembo by Marie Doherty

Printed and bound in the UK
by Clays Ltd, St Ives plc

For Gillian, Chelsea and Rebecca

ABOUT THE AUTHOR

Science writer Brian Clegg studied physics at Cambridge and specialises in making the strangest aspects of the universe accessible to the general reader. He is editor of popularscience.co.uk and has written for newspapers and magazines from *The Times* and *The Wall Street Journal* to *BBC Focus* and *Playboy*. His previous books include *Inflight Science*, *The Universe Inside You*, *Science for Life* and *Big Data*.

Contents

1 Absolute Beginners

<><><><><><><><><><><><><><><><><><><><><><><><><><><><><><><><><><><><>

Relativity is at the heart of this book. We're used to relativity meaning a complex bit of physics dreamed up by Albert Einstein – and Einstein's work is certainly part of it. But there is far more to relativity than that.

To get a feel for relativity at its basic level, we need to take a trip to 1624 to join Galileo on Lake Piediluco in Umbria, central Italy. According to the story, he was being rowed by several oarsmen along the beautiful lake, taking a group of friends on an outing. They were travelling across the water at a good speed by the measure of the day. Galileo is said to have asked one of his friends, Stelluti, if he could borrow a heavy object. Stelluti reluctantly handed over his house key. Four hundred years ago this was not going to be a delicate little Yale key, but was a big iron object – and a one-off that would be hard to replace.

To Stelluti's horror, Galileo took the key from him and hurled it as hard as he could, straight up in the air. The boat, remember, was being powered across the water at a considerable speed. So Stelluti was all ready to leap into the lake, fearing the boat would slip away as the key fell, leaving the precious object behind to drop into the water. His friends had to restrain him, but of course the key neatly dropped back into Galileo's lap.

Whether this story is true is a matter of debate – Galileo accumulated plenty of tales that have little factual evidence to support them. But what certainly was justified was Galileo's confidence in what would become known as relativity. Stelluti had made the very natural assumption that the fast-moving boat would slip out from under the key while the heavy metal object was in the air. However, he hadn't thought through what is truly meant by 'moving'. Galileo had.

In the frame

Relativity comes into play whenever we undertake anything that involves a 'frame of reference' – the specific environment and circumstances in which it is observed – as happened on Galileo's boat. It's both a way of looking at things and an essential requirement to understand how they interact. We use relativity to understand the aspects of physical reality that have no meaning in isolation, but need a frame of reference to give them context. This relativity can involve anything from detecting movement to exploring our place in the universe. Relativity explains how much damage will be caused by a car crash, how we can travel through time, and how gravity does its job. It can be difficult to get a feel for the role of relativity and why it frequently seems to run counter to our common sense expectations: to get a firm grip on this, we are going to build our own universe from scratch.

This is clearly a major undertaking. Realistically, of course, we can only skim the surface of the complexities of the universe. But even so, it will be sufficient to explore the multi-faceted nature of relativity.

The concept of a frame of reference is going to be the central theme in uncovering relativity's role. A frame of reference is the context in which something operates. It can be purely physical. Take a simple statement that you might see in a play script: 'Emma walks from left to right.' Without a frame of reference, we don't know whose viewpoint we are taking. Are we looking at the stage from the audience, or are we at the back of the stage, looking out onto the auditorium? Without a clear frame of reference, we have no idea in which direction the actor playing Emma is walking. So scripts will usually say 'Stage left' or 'Stage right' to make the context clear.

Of themselves, terms like 'left' and 'right' are relative. A frame of reference is needed to make sense of them. Such physical frames give us the most basic form of relativity. So, for instance, when Galileo was on the lake, it was certainly true that the boat was moving when compared with the shore or with the water. That was self-evidently factual. But that movement could not be considered universal. If the boat were moving in the passengers' frame of reference, for example, they would soon be left behind and would suffer a soaking. As far as *they* were concerned, the boat wasn't moving at all. It was the water and the shore – in fact the whole Earth – that was moving backwards for them.

This should have been obvious if, for instance, one of them had put his fingers into the lake. He would feel water moving backwards against his skin. And the same went for the key. *In the boat's frame of reference* the key wasn't moving backwards or forwards, just up and down. So it inevitably fell back into Galileo's lap, rather than being left behind in the boat's wake.

There was a hint of reasoning that made Stelluti's misunderstanding forgivable. Once the key left the boat, the boat and the key had different forces acting on them. Both were being pulled downwards by gravity. Both were being slowed down by air resistance, also known as drag. But the boat also had two other forces at play – the much stronger drag from the water, and the force of the oars pushing it forwards. Given enough time in the air, with nothing to push it forwards, the key *would* have been slowed a little by air resistance and if the key had spent long enough in the air, the boat would eventually have overtaken it. But in practice, for such a heavy object, the impact of air resistance was tiny. If Galileo had thrown a sheet of paper into the air, the result might have been quite different.

However, leaving aside these differing forces, the fact remains that when it was thrown, the key was only moving up and down with respect to the boat. In the boat's frame of reference, it was the Earth that was moving, including the water of the lake, not the boat. Galileo generalised this concept to state that if a boat were moving steadily, and it were totally enclosed with no windows so that it was impossible to see what was happening, and it were insulated from any air movement that could be felt, then there was no physical experiment that could be done inside the boat that would indicate that it was moving.

The human touch

In building a universe from scratch, we need to take in all the physical requirements to make Galileo's relativity possible – and we need to add Einstein's twin works on relativity into the mix, the special and general theories, which include factors that Galileo never considered. However, in trying to understand how human beings fit into the universe, we will have to go further still. If our constructor kit universe is to have humans, it first needs life. And central to the development of life is evolution. Just like stage directions, we can't understand evolution without a frame of reference. Here, though, rather than involving orientation, the reference frame is the environment that makes evolution possible. Evolution is a response to something, whether it is competitors, available resources or even the impact of a DNA reading error producing a mutation. Hence evolution needs a frame of reference, putting relativity at its heart.

When we consider humans, there is one further step to take. We must bring in human creativity, which itself establishes a final type of frame of reference. This is the way we see the world, or the part of it that is involved in a problem we need to solve, an idea we need to generate, or something new we are going to create. Such frames of reference involve relativity just as much as the physical ones, but this is the relativity of *understanding* and *ideas*.

This aspect was highlighted by a famous television show from the 1970s, Jacob Bronowski's *The Ascent of Man*. I still have my parents' copy of the book of the series – the only such book they ever bought. Bronowski, who died shortly after the series was made, was born in Poland and educated at Cambridge after his parents moved to Britain. He spent much of his working life at Cambridge, apart from a period towards the end of his career at the Salk Institute in San Diego, California.

A mathematician who worked in the applied maths field of Operational Research during the Second World War, Bronowski later turned to biology, giving him an unusual breadth of academic experience, which, coupled with a warm yet authoritative personal style, made him an ideal presenter of the series. What made the programmes special was the way Bronowski recognised that it was impossible to separate a history of science from the development of human culture – and the result was a celebration of the breadth of human achievement. As he put it in the book published to accompany the series:

Knowledge in general and science in particular does not consist of abstract but of man-made ideas, all the way from its beginnings to its modern and idiosyncratic models. Therefore the underlying concepts that unlock nature must be shown to arise early and in the simplest cultures of man from his basic and specific faculties. And the development of science which joins them in more and more complex conjunctions must be seen to be equally human: discoveries are made by men, not merely by minds, so that they are alive and charged with individuality.

What *The Ascent of Man* so graphically explored was not the development of science as some abstract, isolated collection of facts. Rather, it established science (and art) as a magnificent flowering that represents the peak of human culture. The title of the series put Bronowski's viewpoint into context. The words are, of course, a play on the title of Darwin's book *The Descent of Man*, but the implications of 'ascent' are clear and unequivocal. We might just be another mammal, in danger of making a mess of a crowded world. We might merely be the inhabitants of a small planet that is nothing more than a speck in a vast universe. But the cultural development that led to the human construct that is science was an impressive achievement.

As Bronowski made clear, science emerges from human culture, and yet it also has shaped and transformed that culture, embedding relativity into our understanding. Modern science can't function without relativity. Frames of reference are essential to make measurements and predictions, to apply physical principles to the world around us. As science has changed our worldview, it has brought relativity to the fore.

Before scientific thinking took a hold, there was an assumption that almost everything around us was based on *absolutes* – ideals and universal truths that humans made efforts to uncover. Yet in reality, so much of nature as we increasingly understand it – from the existence of space and time to the technology that enables us to overcome our biological limits – depends on taking a *relativistic* view.

We are now able to use relativity to develop a wider understanding of our place in the universe, to tell a new version of 'the ascent of man'.

Types and shadows

Early humanity was haunted by that need for absolutes, whether personified in the gods or made philosophical, for example in Plato's doctrine of ideals. This was the notion that there is a pure and absolute reality somewhere out there, but that all we can experience in our human world is a faint reflection of those absolutes. Plato portrayed our existence as shadows, cast from the outer real world into the cave of our understanding.

More poetically known as 'types and shadows', this concept was reinforced in the eighteenth century as Kant's *Ding an sich* (the 'thing itself'), a vision of a kind of absolute reality that we can experience only via what Kant considered human-imposed concepts like time, space and causality. Even such absolutists employ a form of relativity – the relativity of the world we experience to the inaccessible frame of reference of the gods or Plato and Kant's absolute realities. But as we build our universe from scratch in this book, we will see that accessible frames of reference are fundamental requirements of nature. This operates at the basic levels of physics, and as we add in life, the concept of evolution by natural selection will bring in its own need for context and a frame. Similarly, in Bronowski's ascent, it is the reference frames used by the human mind and creativity that enable us to build on our natural capabilities to go further still.

As we will discover in Chapter 8, when we make use of creativity and innovation to produce the technologies that have transformed human existence, it is a result of consciously or unconsciously changing frames of reference. So when we come to put humans in place in the model universe that we are about to create, we need to be aware of the whole edifice of relativity that underlies our position, from basic physical relativity, through the relativistic process of evolution, to the way that human development has set us apart relative to other living things, given our unique* abilities provided by science and technology.

Relativity for beginners

Human beings are inherently relativistic in the way we perceive the world around us. There is a whole business psychology industry built up around relativity in pricing and the way it affects our decisions on whether or not to buy something. Imagine, for instance, you set out to buy a pair of gloves for no more than £20. You see some priced at £40. Ridiculously expensive – you wouldn't consider buying them. Then you see an identical pair at £29.99 and snap them up because they're a bargain … even though they are nearly 50 per cent more than your budget. It was relativity that won you over. The same factor drives the ever-present concept of a sale, where we are impressed not so much by the ticket price but by how much we have saved – even

---

* On this planet, at least.

though the original price might have been an amount that we would never contemplate paying. In the brain, relativity rules.

It is surprising, then, how little effort most of us make to understand relativity, and how rarely it appears in the educational syllabus, even in its Galilean form. Galilean relativity is a powerful yet simple concept. It might seem surprising that it didn't occur to natural philosophers earlier, but it was an uncomfortable fit with the central concepts of cosmology and physics that dated back to the Ancient Greeks and that were only just starting to be questioned in Galileo's day.

It's a mistake to be too blanket-like in describing Greek scientific views. There wasn't a single agreed best approach that lasted throughout the Ancient Greek period. For example, a number of cosmologies were put forward to describe the structure of the universe over a period of 600 years or so. But it was ideas primarily from Aristotle and Plato that were given the most weight in Galileo's time, some 2,000 years later. Largely ignored after the fall of the Roman Empire, the knowledge of the Greeks was rediscovered by Arab scholars, whose translations and commentaries reached the West from the twelfth century onwards. Each of Plato and Aristotle has an important bearing on our story.

As we have seen, Plato's doctrine of ideals provided a universal reality, a fixed point against which the shadows of our everyday existence could be measured. Soon after this was established, Plato's brightest pupil, Aristotle, had firmed up a cosmological picture in which the Earth was the centre of the universe and its position there was fundamental to the behaviour of everything we experienced. This is because his worldview was built on the concept of everything being made of the four elements: earth, air, fire and water. Each of these elements had a natural tendency. Earth and water were influenced by gravity, which meant having a natural desire to be at the centre of the universe. Air and fire were in the grip of levity, which meant that *their* natural tendency was to move away from the centre of the universe.

Add to this Aristotle's notion that apart from these tendencies, things needed to be pushed to keep moving or they naturally stopped (when they were as close as they could reach to their gravity/levity destination), and there was a mindset in place that made it difficult to make the leap to relativity. In Aristotle's universe there were clear absolutes. There was only one centre of the universe and it was uniquely and inevitably the location of the stationary Earth. This fixed the concept of what it meant to be moving. To have an absolute concept of movement you need a fixed, an absolute, reference point, and the Earth as the centre of the universe provided this. So, with an Aristotelian viewpoint, the boat on Lake Piediluco was moving however you looked at it, and required a constant push from the oars to keep it going. The key had no such push, so was going to be left behind. Galileo threw away the misleading absolute fixed Earth, making relative positions and movements the only ones that mattered.

Since Galileo, we have had no such excuse for ignoring relativity. We may teach the basics behind some aspects of Galilean relativity at school, but it is never pulled together into a coherent whole. And it is certainly never identified as being relativity. It was notable that on an edition of the TV show *QI*, the comedian Dara Ó Briain, who has a physics degree, couldn't name Galileo as the originator of physical relativity.

If Galilean relativity is ignored as a concept at school, Einstein's theories of relativity seem to be positively avoided. Their reputation for being incomprehensibly complex puts off any attempt to teach them. Shortly after Einstein published his masterpiece on gravity, the general theory of relativity, the British astrophysicist Arthur Eddington was asked if it were true that only three people in the world understood the theory. Eddington is said to have replied: 'Who is the third?'

This made a good soundbite, but it was a poisonous philosophy that has tainted the way we regard and teach knowledge of the physical world. While it's true that the mathematics of the general theory of relativity was so challenging that Einstein had to get help to understand it, the basic concepts behind his special and general relativity are approachable by anyone. And they should be understood by *everyone*. Yet at the moment we teach physics in schools that mostly dates back to the nineteenth century with only passing acknowledgement of the breakthroughs in knowledge that have occurred since then.

The argument for this approach is that students need to have all the basics of classical physics before they can start to add in the complexities of the key additions of the twentieth century, relativity and quantum theory. And yet that idea comes from a misunderstanding of the purpose of teaching science to children. We don't need to spend the first four or five years of secondary school hammering in the (often tedious) basics of Victorian physics. For the majority who will learn no more science, it's a waste of time that totally destroys the enthusiasm that everyone seems to have for science until the end of primary school. And for the minority who go on to study science in depth, it would be trivial to pick up what is omitted in the basic canon as they go along with more advanced matters.

How much better it would be if we could combine a clearer understanding of what science is and how it is undertaken with more context for where our scientific ideas have come from, based on our current understanding, not a curriculum frozen in the nineteenth century. Again, in our understanding of science, the frame of reference is key. Certainly we should talk about Newtonian mechanics and gravity – but as context for our current theories, rather than all that gets mentioned in any detail.

The omission of relativity, for instance, from secondary school teaching is a terrible mistake, because Galilean relativity is still important in every aspect of life and in the universe. And when we do get on to Einstein, the mathematics doesn't have to be mind-bending. If you decide to risk the Appendix when finishing this book, you will discover that anyone with a GCSE or its equivalent in maths could follow the mathematical argument that shows that time travel is possible. How much more exciting to have been taught *that* in school physics than calculating the work involved in pushing a block up a slope.

◇◇◇◇◇

Throughout this book we will explore how relativity is intertwined with the effective development of an understanding of the place of humanity in the universe. We will get a better feeling for how the basic components of the universe work – and how remarkable both life and human creativity are.

To get a clear picture, we are going to build and populate a virtual universe, step by step, adding the layers necessary to end up with the scientific and technological achievements of human culture. We will need material to build our universe, time and movement, forces, notably gravity, to assemble our basic building blocks, the development of life and the human ascent powered by creativity and science. But before we can add anything, we need to get to grips with the unnervingly slippery topic of empty space.

2 Space

Over the next few chapters we are going to undertake a dramatic experiment. The plan is to construct a universe from scratch, up to and including human inhabitants. This is, of course, just a *thought* experiment – no version of reality will be harmed in the process – but it will still require some creative work.

The first requirement on our ingredients list is space. Like many of the constituents of the universe, space is something of which we have an inherent grasp, yet still find hard to describe. We think of space as a kind of container, a three-dimensional emptiness which provides the context for everything physical that will populate the universe. (Those three dimensions are an assumption we will need to test a little later, but it will do for our initial conception.)

As yet, in our universe construction kit, space is the single unique ingredient, so we are dealing with true and absolute emptiness, something that will never be able to exist once the construction of our universe is complete. Space alone is a total, everything-spanning nothingness. This is inevitably hard to visualise. We have no experience of truly empty space. In our everyday lives, we spend our time surrounded by things, by movement, by the relentless tick and tock of time. Even if we envisaged going out into the depths of space (it's unfortunate that we don't have a scientifically acceptable separate term for what used to be called 'the heavens'), we wouldn't experience true emptiness. There is always dust, always light crossing that space from other objects. And simply by being there we ensure that the space isn't truly empty.

In reality, it is probably impossible to get the mind around pure and absolute emptiness in a satisfactory fashion. We're used to hearing about the concept of infinity as something that is beyond true human conception, but it can come as something of a surprise to discover that we also have a titanic struggle to envisage total emptiness. In this limitless expanse of empty space there is no frame of reference, nothing with which to pin anything down. Here we have a true absolute – the absolute absence of *anything* material. Relativity is impossible in our starter universe of pure space because this is an empty unity. Relativity implies a relationship, and a relationship needs more than a singular entity. So far, our featureless universe is the ultimate solipsist.

This impossibility of establishing relativity in emptiness becomes more obvious once we consider the language that is necessary to deal with familiar relativistic concepts. It is important to remember that what is meant by 'relativity' at this basic spatial level is the simple Galilean view, typified by his experiment (or prank) in the boat on Lake Piediluco.

The featureless void

With Galileo's picture of relativity in mind, we can discard for the moment exotic conceits like boats and people, keys and lakes and movement, to rejoin our empty, featureless space. Here we discover that any attempt to introduce relativity is littered with terms like 'with respect to' or 'in this frame of reference'. If I'm moving at 50 kilometres per hour (kph), for example, there is an immediate question we need to ask (let's not worry too much about the concept of 'I' or how we measure hours in an empty universe at the moment – this is just a thought experiment). I am moving at 50 kph with respect to what?

In our everyday lives this doesn't seem a problem because, like Aristotle, we habitually think of 'stationary' as being defined by the Earth – the sphere of our world forms our default frame of reference. So if I say that I'm driving a car at 50 kph, it's inevitably assumed that I mean I'm moving at that speed with respect to the ground. But that *is* an assumption.

If I'm in a collision with another car, what's important is my speed with respect to that other car, which could be totally different depending on whether we're moving in the same direction or in opposite directions. If the other car is just ahead and moving in the same direction as me at 49 kph, I crawl towards it at just one kilometre per hour, with no real damage caused on impact. If the other car is heading towards me at 100 kph, its speed in *my* frame of reference is 150 kph, the result of adding our speeds together, making for a horrendous crash. (My speed in its frame of reference is also 150 kph, but heading in the opposite direction.)

That's why we can't manage without relativity in the normal world. It is our relative speed that determines the outcome of the collision, not some arbitrary speed in the reference frame of the Earth. But what about our empty universe? If I suddenly appeared in that universe as a unique observer, unless the universe has detectable boundaries (which arguably would stop it from being empty), there is nothing with respect to which I can measure movement. My only frame of reference is myself, and I can never be moving with respect to myself. There is no useful relativity.

For Isaac Newton, writing around 60 years after Galileo's lake trip, the need for an external frame of reference was clear. Space itself, he believed, was an absolute concept that was 'homogeneous and immovable'. This he contrasted with the relative space that was the result of measurement. Clearly such relative space is impossible for the moment in our model universe, because there is nothing present to measure (or, for that matter, to make a measurement with). Newton used the examples of objects moving through absolute space to obtain relative space. He pointed out that the air around the Earth occupies the same relative space when compared with the position of the Earth as the Earth moves around the Sun. That air is constantly changing the position that it occupies in absolute space, but we can't give a measurement for where that absolute position is.

Getting a grip on the nature of space is an essential if we are to build ourselves a universe. It's something that creation myths can take for granted, but that is impossible if we are to take a scientific view of the universe and its origins. Traditionally our understanding of space has been as a continuum, something that can be divided up in whatever way we like. But this understanding is challenged by quantum theorists.

The 'quantum' in quantum physics refers to a piece of something that comes in minimum sized units. During the twentieth century, it became clear that many apparently continuous phenomena, like light, were actually quantised, coming in tiny chunks. Most quantum physicists believe that space itself is quantised. It's as if it were like a jar of salt, rather than a jar of water, having distinct 'grains', although on an extremely tiny scale. If this is the case, it may be that the granular structure of space gives us the potential for some kind of frame of reference. We also need to consider whether the space in our model universe is infinite or finite. Does space have an edge or a centre that could give us a signpost to make a kind of relativity available?

Nature abhors a vacuum

The notion of truly empty space fascinated and horrified early thinkers. Much Greek argument on the subject, notably that of Aristotle and his followers, was based on variants of the idea that nature 'abhorred' a vacuum and as such, truly empty space was meaningless. (It's more accurate to call what the Greeks considered abhorrent a 'void' rather than a vacuum, as you can have a vacuum but still have, for instance, gravity – but they were referring to a total absence of everything.) One of Aristotle's arguments against the existence of an empty void entertainingly used what would become Newton's first law of motion. This was because, if there were a void, Newton's first law would have to apply, and Aristotle thought that this was self-evidently wrong.

Aristotle argued in his book simply titled *Physics* that in such a void, 'no one could say why something moved will come to rest somewhere; why should it do so here rather than there? Hence it will either remain at rest or must move on to infinity unless something stronger hinders it.' The argument that Aristotle uses (meaning it to be something clearly not true) has a remarkable similarity to Newton's first law, which was originally stated as: 'Every body perseveres in its state of being at rest or of moving uniformly straight forward, except insofar as it is compelled to change its state by forces impressed.'

For Aristotle it seemed obvious that there were two aspects to motion. As we have seen, in his philosophy, things had a natural tendency to head for where they 'belonged' – earth and water moved towards the centre of the universe, while fire and air headed away. Apart from this, things had a natural tendency to stop unless they were pushed. This was a direct result of observation. Things that are moving do, on the whole, stop unless we push them. Especially if things move relatively slowly and your technology can't produce low-friction bearings. So, for instance, a wooden cart will stop very quickly if left to its own devices. But what Newton saw (as did Galileo before him) was that underlying this apparent nature of motion was something more fundamental, a tendency that was being interfered with by gravity, by air resistance and by friction.

Of course in our empty universe we are yet to have any object that can move; nor do we have gravity, air resistance or friction to influence that motion. But the irony is that Aristotle, in contemplating the void, came up with what was arguably one of the best scientific observations he ever made: that if there were a true void with no influences, no forces in action, an object would stay still, or remain forever in continuous motion. And as, for Aristotle, this clearly never happened, he considered it a useful argument to show that empty space, a void, could never exist.

The thirteenth-century natural philosopher friar Roger Bacon used the sheer emptiness of the void to argue that there cannot be a vacuum between us and the heavens (i.e. where the planets and stars are), because if there were, he believed that we couldn't see the light that came from them. He wrote: 'In a vacuum nature does not exist. For vacuum rightly conceived is merely a mathematical quantity extended in three dimensions, existing per se without heat and cold, soft and hard, rare and dense, and without any natural quality, merely occupying space.'

And without the very existence of 'nature' he could see no mechanism for light, which he believed moved by continuous interaction with a medium in a process known as 'multiplication of species', to get from one place to another.

Bacon would be proved wrong about the ability of empty space to prevent light passing through it. But our 'space' is not a true void, flagging up an interesting observation that makes cosmologists' most frequent speculation about the beginnings of the universe feel like it's on uncertain ground. Quantum theory predicts that a vacuum will not be empty, but will instead seethe with virtual particles that briefly pop into existence, then disappear again. The very beginning of everything is sometimes represented as such a quantum fluctuation where something briefly pops into existence out of nothing, but then is influenced by other processes to inflate into a universe with contents that will no longer briefly exist, then disappear. This way, it seems it is possible to start with an entirely empty universe and to get to something that could eventually become the whole, complex, well-populated universe we observe today.

However, the sleight of hand involved in creating this sophisticated physics model covers up a gaping hole. Let's think about our model universe with nothing whatsoever in it. At this stage, just as Bacon describes, it is simply a mathematical extension of dimensionality in three perpendicular directions. A true void. Absolute nothingness. Where are the physical laws and universal constants? What embodies those laws? Where do they come from? What brings them into being? What tells virtual particles to pop into existence in what should have been a totally empty space? It seems that this can't be a true void after all.

Cosmologists agree that there could be universes with physical laws and universal constants that are different from those we experience. Or that could have no universal constants at all. So it isn't enough to say that somehow the very existence of space alone is enough to call the laws and constants into being. (And to do that immediately throws away the idea that we are starting with a true empty space.) Yet as soon as we accept that the laws and constants are not an unavoidable consequence of the existence of space, but could have a range of values, then they become an add-on. They are an addition to the void.

The natural laws

Let's take a moment to consider what we mean by the laws of nature and universal constants, as we need to know just what it is that is being added to our empty void when we construct a universe. The great twentieth-century physicist Richard Feynman described physical laws as follows: 'There is … a rhythm and a pattern between the phenomena of nature which is not apparent to the eye, but only to the eye of analysis; and it is these patterns which we call Physical Laws.'

I have to be honest here – I dislike the term 'law' being used in science as it implies something fixed by statute, something that is agreed and binding. But what we call a natural or physical law is both stronger and weaker than the laws that are used in courts. Physical 'law' is stronger than a legal law, because it isn't made up by human beings and doesn't need our acceptance to be operated. It exists whether or not we agree with it, and whether or not we are there to observe it, codify it and apply it. Yet physical law is also weaker than a product of our legal system, because the kind of laws that are applied in court are written down in black and white. You can read the statute and while you can quibble about the interpretation of it, you can't argue about what the words are. Those words are *fact*.

Unfortunately, despite the way the term is used in poetic descriptions of science, there is no 'book of nature'. You can't just go and read the laws and discover exactly what they are. They are deductions (or, more accurately, inductions) from the best evidence we currently have, always subject to revision. They are never truly *fact*. As Feynman said, the laws are discovered by analysis – not defined by decree. He uses an example of a game of chequers, where the fundamental laws are the rules by which the pieces are allowed to move. Watching a game, we can use logic and mathematics to establish a best guess of what the rules are, but we can't arbitrarily state what they are. Like the inhabitants of Plato's cave, we can't actually see the laws in their pure form, we need to deduce what the law on the outside is from the shadows we observe via our limited view.

When we 'discover' a natural law, it's a bit like having a black box with a number of openings on its sides, into which we can drop a ball bearing. Inside the box is a structure. It is a structure that is real, but that we can never observe. By putting a ball into each of the openings and moving the box about so that we can hear its movements, and by observing which opening the ball bearing eventually comes out from, we can make some deductions about the structure in the box. If we have a highly sensitive metal detector, we can probably make even better deductions by following the ball on its path.

The mental picture we build of the structure inside the black box, which scientists would call a 'model', is like our formulation of a natural law. Our model of the structure in the box may well be incomplete – there could be a part of the structure the marble never touched. Or the marble might be too big to go into every nook and cranny, so we might then miss the fine detail of the structure. Later on, we might come up with a much smaller ball bearing as our research tool, which would enable us to get into more of those nooks and crannies. These new results don't dismiss the research based on the larger marble, but they give us a better model of the 'law' of the inner structure.

We can see the same thing happening with Newton's laws of motion and his law of gravitation, both of which we will meet later in the book, once we have added enough to our universe to be able to deal with movement and gravity. In each case, Einstein later came up with relativistic versions, which provide the ability to get further into the nooks and crannies of these physical laws. Newton's version was a good approximation and all we need for most everyday uses, but Einstein's version provided much more detail.

As the black box and marble model makes clear, however, we can never be certain that our descriptions of the natural laws are perfect and complete. There is always the opportunity for a different kind of mental or experimental probe that will result in the discovery of new and amazing nooks and crannies. We could even discover that what we thought were a series of walls within the box were actually indentations, or thin bars. With a different approach a probe might pass straight through them. So we always need to remember that we don't really *know* the natural laws – we know the approximations that our current tools, both experimental and mathematical, enable us to put together.

The universal textbook

The Nobel Prize-winning physicist Steven Weinberg, when interviewed about his career, said that he was driven by a desire to contribute to the 'ultimate textbook', a hypothetical book which contained in its first chapter a few principles that were the closest we could *ever* come to the ultimate laws of nature. Rather like the axioms used in mathematics as starting points to construct theorems in a logical, step-by-step fashion, this 'chapter one of the book of everything' would give us the laws we needed to construct all of science – certainly all of physics.

It might seem that the world would be a more boring place if Weinberg's chapter one had already been written. When Max Planck, the physicist who would turn our understanding of nature on its head by kickstarting quantum theory, was at university, he hesitated between a career in physics or in music, as he was an accomplished pianist. His physics professor, Philipp von Jolly, told him to go for music, as there was little left to accomplish in physics. All that was left to be done was to fill in the small details.

Thankfully, Planck ignored von Jolly and went into physics anyway. But Weinberg argues that the advice was based on a false premise that science would be limited by knowing the basics perfectly. Weinberg draws an analogy with the early maps. 'In the middle ages Europeans drew maps of the world in which there were all kinds of exciting things like dragons in unknown territories.' But, Weinberg says, we are better off knowing the fundamentals – that dragons don't exist – and being able to work instead on the interesting detail. The first chapter of the great book might give us all the basics, Weinberg argues, but it's by building on them that we make life interesting, just as a dictionary and a book of grammar give us the basics of writing, but it's what a writer can do with these components that matters.

Yet all the evidence is that, in reality, this vision of perfection that has been Weinberg's driving force is a mirage, much like von Jolly's imagined near-complete view of physics. Yes, we might be able to simplify laws. And, yes, we might be able to use smaller and smaller marbles to get a more precise match with reality. But we are never going to see inside the black box. And there will always be the possibility that a new approach will open up whole new expanses of the box that totally transform (whether to simplify or make far more complex) our picture of what is inside. Weinberg has been seduced by mathematics, where such perfection is possible because mathematical laws are like legal laws. In maths, *we* decide what the axioms (effectively the basic mathematical laws) are. There is no such possibility in the physical world.

The most fundamental of the natural 'laws' have no reasoning behind them – they are purely the result of observation. Take one of the simplest, the law of inertia. This was discovered by Galileo when rolling balls along inclined planes, sloping bits of wood with a channel to keep the ball from falling off the edge. Galileo found that (unsurprisingly) if he rolled a ball downhill, it got faster. Similarly, if he rolled a ball uphill, it got slower. But the clever part was to make the leap of understanding and point out that it only seemed logical that when rolling a ball on the flat, it would neither get slower nor faster, but would continue to roll at the same speed, unless something interfered with it.

Galileo's discovery became incorporated, with some fancier wording, as Newton's first law of motion. As we have seen, Isaac rendered this as, 'Every body perseveres in its state of being at rest or of moving uniformly straight forward, except insofar as it is compelled to change its state by forces impressed.' Or, in a more modern wording, a body will remain at rest or in steady motion unless it is acted on by a force. This doesn't seem a natural observation, because everything we experience in everyday life is already acted on by forces – by friction and by air resistance, for instance – and these tend to stop something that is moving. But without those forces in play, the movement would continue for ever.

Without Galileo's experiments (and those of others who followed him) this would have been a huge mental leap. Now we can see something like this happening in space, where objects do pretty much keep moving indefinitely once they have begun to move, if not acted on by gravity or impact. And all the evidence we have is that the law of inertia is true. What we don't know and can't say is *why* this happens. It just does. It is part of nature – part of the something that we assume was there when the universe began. Part of the underlying matrix of reality that means that even apparently empty space contains something. Equally, we can't say for certain that there aren't circumstances where it's not true. We assume it is universal for convenience, but there is no way to prove this.

Some of the other physical laws are one step removed from being a reflection of a fundamental aspect of the nature of the universe. Given basics like the law of inertia (the concept 'inertia', incidentally, sounds sophisticated and scientific, but it only means 'having the properties described by Newton's first law'), it is possible to construct these extra laws. So they aren't fundamental, but scientists have found it useful to consider them as laws in their own right to avoid having to rebuild them every time they are used.

Pragmatic simplicity

This importance of usefulness is reflected in another comment that Feynman made on the subject of natural laws, using gravity as his example: 'But the most impressive fact is that gravity is simple. It is simple to state the principles completely and not have left any vagueness for anybody to change the ideas of the law. It is simple and therefore it is beautiful ... This is common to all our laws; they all turn out to be simple things, although complex in their actual actions.'

Note that Feynman is not suggesting that this means that working out the details of what will happen as a result of gravity is simple. As we will see when getting on to the equations that lie at the heart of the general theory of relativity (see page 169), there is mathematical complexity here that caused even Einstein to struggle. And much of the basics of, say, quantum theory seems crazy. The natural laws are not necessarily well represented by common sense. Nevertheless, I can explain the principles of relativity or quantum theory to primary school children and they can grasp what is happening (better, arguably, than some adults). They can't do the maths, but they understand those simple principles.

This requirement for simplicity illustrates a problem with modern theoretical physics, which is largely driven by complex mathematics, rather than by readily grasped principles. It is no surprise, for example, that everyone from scientists to news reporters struggled to explain the significance of the Higgs boson to the world, when it was (probably) discovered at CERN in 2013. Some physicists argue that we have gone too far in our over-dependence on building complex mathematical structures. It may be we will see these concepts thrown off when a new, simpler underlying set of laws is discovered. Time will tell. But there appears to be some kind of framework of reality that we interpret using the human approximations of our physical laws.

As for the constants that are part of the foundations of nature, these are effectively fixed components of the mathematical models we make to represent natural laws. A constant is simply a numerical value that does not change with time. Some constants are very useful, but are local and not relevant to natural laws. So, for instance, it's handy to know that there is always a bus from my nearest stop on the hour. It's a constant, but it's not exactly a fundamental aspect of nature.

Similarly, the forks I use when I set the table are always the same length (give or take manufacturing error and a spot of expansion and contraction due to heat), but the value that specifies their length is of no use to science. These kind of local constants can have a relatively short lifetime as well – the bus company could change its timetable, or I could change to a new set of cutlery. However, the constants that are of interest in science are mostly 'universal', which suggests both that they apply anywhere in the universe and anywhere in time (each of these is an assumption, which we'll explore in a moment).

Let's take an example. The speed of light is probably one of the best-known universal physical constants. To be precise (and we have to be precise in science), the constant in question is the speed of light in a vacuum – light gets slower when it passes through a transparent medium like glass or water. We usually represent the speed of light as 300,000 kilometres per second, or 186,000 miles per second, which are handy approximations, but the value is actually 299,792,458 metres per second. Unlike many values for constants, this is not just the best value we have from current measurement, which will change when we can undertake more accurate experiments – it is a definitive and exact number.

The reason for this being the case is scientific pragmatism – the metre is defined as 1/299,792,458th of the distance that light travels in one second. So although the exact length of a metre will change over time, as better measurements become available, the speed of light won't. (It's a shame that this approach wasn't settled on until the metre had already been defined very accurately. Otherwise, there is no reason why the speed of light could not have been *defined* as 300,000,000 metres per second, which would have made calculations and remembering the exact speed a whole lot easier.)

Knowing the value of the speed of light, and the assumption that it is unlikely to change, does not make it a value that is worth calling a universal constant in its own right. We can measure all kinds of values that don't necessarily change with location or time. But some such values crop up frequently in the patterns that Feynman mentioned as being the basis for the natural laws. We see this happening with the speed of light in what is probably the best-known equation ever – an equation that is relativistic to its core:

$$E = mc^2$$

The c in the equation is the speed of light.* As it happens, there is an obvious connection here, in the sense that information about light was involved in the derivation of this equation. But this isn't the case with all universal constants. For example, the constant that emerged from Newton's work on gravity, G,

---

* If you've ever wondered why c is the speed of light rather than l for light, s for speed or v for velocity, it's not entirely clear. Einstein did initially use v, following the lead of the developer of electromagnetic theory, Maxwell. But this gets confusing when comparing the velocity of light with that of a moving body. The earliest known use of c explained it as c for constant, but it has also been identified as c for *celeritas*, the Latin for speed.

is not a measurement of anything – it is simply a consequence of the pattern of the observed effects of gravity. It is the number that, when plugged into the equation, happens to work. Other constants crop up so frequently that, while they are clearly very important in nature, it can be difficult to understand why they are present in a particular equation. Let's take a look at an example featuring the well-known universal mathematical constant pi (π).

Why pi?

The physicist Eugene Wigner, when writing about the unreasonable effectiveness of mathematics in describing reality, told a story of two high school friends who were discussing their careers. One of them, a statistician, was talking about his work. The statistician showed his friend a paper describing the way populations change with time. He told how a particular curve, the Gaussian distribution, made it possible to predict the behaviour of those kinds of population.

The friend wasn't impressed. He couldn't see how the statistician could possibly know that the graph he drew, a particular shape that emerged purely from the maths, could somehow predict the way a group of living, thinking organisms would behave. But it turned out there was something worse hidden among the hieroglyphics. The friend pointed to a symbol and asked what it meant. 'That's pi', said the statistician. 'You know what that is. The ratio of the circumference of a circle to its diameter.' The friend gave a rueful smile. 'Now I know you're messing with me. What has the population got to do with the circumference of a circle?'

Universal constants can be like that – they creep into all

manner of calculations where it really isn't obvious just why they have cropped up, without delving into a lot of analysis. Pretty well all of science, with a few daring exceptions, is based on the assumption that such constants are universal in time and space. So it doesn't matter when or where you are – as long as you are in the same universe, the constants will be the same.

Initially, this assumption was almost entirely one of convenience. In the case of the physical constants like the speed of light, there is no reason why the speed has to be the same in all cases (mathematical constants like pi have more justification because they are defined from abstracts rather than experiment). But having a variable constant makes it very difficult to ever come up with a simple scientific explanation, as the ground would be constantly shifting under the feet. If constants varied randomly with time or location, science would become pretty well impossible to perform.

There is now good evidence that many constants have not varied much with time. The electrical charge, for instance, can be traced back at least a couple of billion years by using the remains of natural nuclear reactors, where a uranium chain reaction was started in the distant past without human intervention. Measurements taken on the remains show that nuclear reactions took place that are very sensitive to the electrical charge, and from the measurements of its effect, that charge has stayed pretty much the same during that period. Other scientists are looking even further back in time, making use of the way that space acts as a kind of time tunnel. As light takes time to reach us, the further away we look out into space, the further back in time we look. (Measuring how far back we are looking relies, of course, on the assumption that the speed of light has remained constant.)

A fine constant

Perhaps the most remarkable experiment to attempt to explore the consistency of the universal constants is one undertaken by astronomers at the University of New South Wales and Swinburne University Technical College, both in Australia. This experiment has used two of the largest telescopes in existence, the Keck telescope in Hawaii and the Very Large Telescope in Chile, to peer out to distant quasars, radiation sources so far away that the light has taken about 10 billion years to reach us. A quasar is thought to be the radiation generated by matter falling into a supermassive black hole at the heart of a distant, ancient galaxy.

The experiment was not observing the quasars themselves, but used them as backlights to study the absorption of the light by intervening material. When light passes through matter, the matter tends to absorb certain colours, leaving dark 'absorption lines' in the colour spectrum. This is how we can identify which elements are present in stars. The distance between key absorption lines in the spectrum is dependent on one of the universal constants, a value known as the fine structure constant.

In reality, the fine structure constant is more of an amalgam of other fundamental constants – usually represented by α (alpha), it is proportional to e^2/hc, where e is the charge on the electron, h is Planck's constant which links the energy in a photon of light to its colour, and c is the speed of light. The fine structure constant has the advantage of being dimensionless – it is just a number, whereas, for instance, the speed of light has the units of metres per second. This lack of dimensions helps because it means it's possible to detect changes in α where an apparent change in the speed of light, say, could be the result of

31

a change in the electrical charge, because the definitions of the constants depend on how their units are defined. (Remember that a metre is defined in terms of the speed of light.) Lacking units, the fine structure constant, which for no good reason is very close to the number 1/137, does not suffer from this problem.

At the time of writing, although the results are not conclusive, there is reasonably good evidence that α has indeed varied by a small amount over billions of years, although this variation was different depending on the direction the observers looked out into the universe. Further research is needed, but if the variation holds up, it could have very significant implications for many of our basic assumptions about the current cosmological model, which requires constants like this not to change at all.

As well as taking for granted that universal constants do not vary in time, it is also an unproved, if convenient, assumption that they do not vary with circumstance. Take, for instance, the gravitational constant G, which defines how strong the attraction of the gravitational force will be between two objects given their mass and separation. This is considered to be the same for electrons as it is for apples as it is for planets as it is for galaxies. However, once more, what we have here has no experimental basis.

We do know that not everything behaves in the same way at different scales. Atoms and other quantum particles do not generally behave like 'macro' objects like apples and planets. They are dependent on quantum mechanics, acting as if they don't have true locations except when undergoing interactions – in the end, very different from the everyday objects that they make up. Yet we blithely assume that, for instance, the way that gravity and Newton's laws work on the scale of the solar system also applies exactly to something as large as a galaxy.

If this is the case, there's a problem. When things spin around (and pretty well every body in the universe does – see page 114), there is a natural tendency for parts of the spinning object to carry on in a straight line and fly off, rather than stay in the body. It's only the gravity of the body, or the electromagnetic force holding a solid together, that stops them from doing so, pulling the parts into the spin rather than allowing them to fly off. But it has been known for some time that galaxies, for instance, survive spinning at sufficient speed that physics predicts would cause large parts of them to become detached.

The current favoured explanation for this is that there is an extra kind of matter called dark matter, of which more in the next chapter (page 61), a substance that is supposed not to be influenced by electromagnetism – so we can't see it or touch it – but does have an influence gravitationally. If there is enough of this stuff, appropriately distributed within a galaxy, then it would provide enough attraction to hold the whole thing together, and would explain a number of other behaviours in the dynamics of galaxies and clusters of galaxies.

The only problem is that to explain what is observed, there has to be a whole lot of this never-yet-detected stuff out there. Over five times as much as there is ordinary matter. That's a big fix to our model to account for the odd behaviour of large bodies in space. Some have suggested instead that what we are seeing is a phenomenon of a universe where some of the laws and constants don't apply in quite the same way they do with familiar everyday objects when applied to the scale of a galaxy. This approach, making small modifications to Newton's gravitational predictions, called 'Modified Newtonian Dynamics' or MOND, is not perfect – it does not explain all the effects ascribed to dark matter – but then current dark matter theories don't explain all current observations either.

The MOND theory is in many ways the simpler explanation of the two, but one that many physicists are reluctant to even contemplate, perhaps in part because of the need to give up on the universality of a constant. It makes sense, as usual in science, to go with the best-accepted theory until there is sufficient evidence to make a change necessary, so for the DIY universe we are constructing, we will consider it a requirement to have dark matter present, but with the proviso in mind that it may not be the ideal solution for the real universe.

Which way is up?

Returning to our simple universe model, this is still, as yet, just empty space. Even if we do allow the existence of some physical laws (which we don't), there is surely another unfounded assumption in the picture of the beginning of the populated universe as a result of quantum fluctuations. Our truly empty three dimensions of space are just that. There is no time as yet. But how is it possible to have virtual particles popping in and out of existence in quantum fluctuations in a timeless space? We are a long way yet from having a workable universe.

Before we move on to add in some components, to make those three dimensions a little less lonely, there is one final consideration to make. Why should we choose three dimensions? Where did that number come from? In a mathematical sense there is nothing unique about three dimensions, though three dimensions do have some special properties that make them desirable as a minimum.

For mathematicians, any number of dimensions can be considered for a space – even, with a certain amount of mind-twisting, fractional dimensions. While it's pretty well impossible to envisage, say, 50-dimensional space as a real space, it's perfectly possible to keep adding extra dimensions mathematically without any limit. It can sometimes be useful, for instance, to have a virtual multi-dimensional space in which every possible value of a property is represented by a different dimension. This space isn't 'real' but it has practical value for calculations.

Among mathematicians there is a certain kudos in dreaming up structures in vast numbers of dimensions. A favourite with a certain kind of mathematician is something called the 'Monster group', which is a mechanism for reflecting the different ways something could be rotated if you had 196,883-dimensional space available (this particular number results in some mathematically interesting properties). However, while such multi-dimensional imaginary space can be valuable in performing calculations, no one realistically suggests that the universe has vast numbers of spatial dimensions.

We know that the number three for spatial dimensions reflects our experience. We can move up/down, left/right and back/front; three dimensions of movement are sufficient to take in all of known space. Each of these dimensions can be placed at right angles to all the others, and then we run out of new directions to go in the universe we experience.

Having at least three dimensions is an essential for the realistic existence of life – certainly life as we know it. In 1884, an English head teacher called Edwin Abbott (to be precise, Edwin Abbott Abbott) wrote a slim book called *Flatland* in which he described the rather dull adventures of creatures mostly living in a two-dimensional world. Abbott was one of the first to think through the implications of living in different dimensional spaces.

As an obvious example of the problems a two-dimensional entity would face, it could not have a digestive system like ours with separate entry and exit points, because as soon as you link two openings on a two-dimensional body, the body is cut into two separate parts. A third dimension is necessary for two openings to be linked in a single body.

Some physical theories require that there are extra dimensions over and above the familiar three, dimensions that are either curled up so small that we can't detect them, or that are effectively external to our universe, so the universe we are familiar with makes up a three-dimensional membrane (or 'brane') floating in extra-dimensional space. As yet there is no experimental evidence to support these theories. They simply work in the mathematics.

However, there is one way that we can invoke an inaccessible fourth dimension that would be consistent with observation and that could be potentially useful. This has an influence on the extent of space. The space in our DIY universe could be infinite, stretching in all directions for ever. And this could be the case in the actual universe; we have no way of knowing. We can see only as far as light has had a chance to travel, which, given theories on the expansion of the universe, is probably about 45 billion light years in any direction. But whether the universe is finite or infinite beyond that is not clear.

While an infinite universe has certain philosophical attractions, we tend to raise an eyebrow at anything physical that embodies infinity. Part of the problem is that so much of our understanding of reality depends on mathematics, and mathematics struggles with actual infinity, as opposed to infinity being a never-reached limit, as it is used in calculus. Infinity is not a number in the normal sense. It does not obey the usual rules of arithmetic. Infinity plus 1, for example, is just infinity. So although we can't dismiss the possibility of an infinite universe, it is often seen as convenient to have a mechanism to design a finite universe.

Thinking purely in three dimensions, if our universe has limits there are some distinct problems. What happens at the edge? What lies beyond the edge of everything? What ideally we want, if our universe is to be finite, is a mechanism for it to be finite but not to have boundaries. And there is a mechanism to make this possible, based on the model of a similar effect in two-dimensional space. It's a situation we are very familiar with – the surface of the Earth.

If we ignore oceans, the surface of the Earth has some very interesting properties. It is finite, certainly. Yet we can walk in any direction for ever and never reach the edge of the planet. This is because the apparently two-dimensional space of the surface of the Earth is, in reality, folded in a third dimension so that anywhere we would expect there to be an edge we simply re-join the surface from the other side.

Extending this idea to a three-dimensional space, we could have a finite but unbounded universe if the universe folded back in on itself thanks to an unreachable fourth dimension, so that heading out of the universe in any direction would result in heading back into it from the opposite direction. Just like on the surface of the Earth, there would be no way out, despite the universe being a finite entity. There have been suggestions in the past of astronomical evidence that hints at this happening. Close to the edge of the observable universe there may be structures on one side that can be seen from the opposite direction. But, as yet, no such evidence has stood up to rigorous assessment – and even if we did live in such a universe, there is no reason why the effect would have to be visible in the observable universe.

◇◇◇◇◇

This is pretty much as far as we can get with building our toy universe based on space alone. We have, after all, a universe that is as yet very boring. But things get a lot more promising – and relativity has the chance to blossom – once we add in stuff.

3 Stuff

<<<<<<<<<<<<<<<<<<<<<<<<<<<<<<<<<<<<<<<<<<<<<<<<<<<<<<<<<<

With stuff – things, objects, whatever you want to call it – space takes on a new, more navigable depth, as long as that navigation is mental and instantaneous – true navigation will also require time. That wholly relative concept, position, now comes into play. Any single particle gives a frame of reference with which to locate another one. And once we have two or more particles we can compare characteristics like mass.

It's important that we use the apparently unscientific term 'stuff' here, because the more proper sounding 'matter' isn't sufficient to cover what is being added to that empty space. After all, light, for instance, is not matter. And yet it is a hugely significant part of the 'stuff' that will enable space to be more than an empty void. We will also discover that 'stuff' brings into play fundamental forces to add to the complexity of our universe.

At the same time, stuff gives us what is arguably our second absolute. The fundamental particles that constitute stuff and three of the forces that handle most interactions of stuff (see page 56; we will come back to gravity in a separate chapter) seem to be invariant. Yet despite their independent nature, they still have strange linkages. There is no obvious reason why, for instance, a proton (or more precisely the sum of the quarks that make it up) and an electron have equal and opposite charges, yet it surely can't be a coincidence. Is there any reason why there are two and only two opposite charges, negative and positive? And do we really need a Higgs boson? Stuff both brings relativity to space and opens up its own mysteries.

We experience matter and light all the time, even though our everyday interaction with matter is, in effect, a constant lie, as we are fooled by the collective behaviour of billions of tiny particles, which examined individually act in ways that seem contrary to nature.

Once we are dealing with stuff, space takes on a far greater significance, bringing in the potential for a much wider range of frames of reference. Without space, all matter would exist in a single location, which would be inconvenient to say the least. There could be change in such a single-location universe once we had time (coming in the next chapter), as bits of matter winked in and out of existing, but nothing very meaningful could take place.

It's also true that our ability to deal with the world around us is predicated on the benefits of having plenty of space for stuff to operate in. For example, the universe is full of risks, and we are largely risk-averse organisms. Yet if we had to take into account every risk that exists in the universe – from a black hole located a few million light years away, or an out-of-control car on a different continent, for instance – we would be incapable of taking any action because the sheer number of risks we faced would be overwhelming. In practice, though, the vast majority of such risks are too far away to be a concern, and we are left with the relatively few threats that are spatially imminent. The expanse of space protects us from risk overload.*

Elementary matters

If we are to introduce stuff to our model universe, we need to know what is going to be added to the void. As already mentioned, there does seem to be a kind of absolute quality to the nature of stuff – to the components of which it is made. Take a zoom in to the most familiar aspect of stuff, matter, and you will find a relatively small number of elements: around 94 of them.**

Around 94 is a small number when you consider just how many atoms there are in the universe. Clearly we can't put an exact value to that number, but we can make an approximation to the number of atoms in a star, how many stars there are in a galaxy and how many galaxies are in the known universe. With a fudge factor for all the bitty amounts of stuff that occur outside of stars (bear in mind that stars are big – the Sun contains over 99 per cent of the mass of the solar system), it has been estimated that there are around 10^{80} atoms in the observable universe. That's 1 with 80 zeroes after it. And every single one of these is chosen from fewer than 100 kinds of atom.

---

* Arguably, this is one of the reasons that today's world seems such a complex and difficult place. A few hundred years ago, most of us had only a very localised experience or awareness of space.

** It used to be thought that uranium, element 92, was the heaviest naturally occurring element, but there are natural fission reactors that have resulted in plutonium being produced. At the time of writing, the periodic table includes a total of 118 elements, but everything up above plutonium is something of a fake element, an unnatural construct with a ridiculously short life.

With a touch of school science, we know that at heart, things are even simpler than that. Each of those atoms – from the simplest, hydrogen, up to those messy artificial elements – can be assembled from a constructor set of just three particles: neutrons, protons and electrons. The relatively massive neutrons and protons form a small central core, while the electrons occupy the outer part of the atom in a fuzzy cloud of probability.

That, when you come to think of it, is a truly remarkable case where universality really does seem to apply. Matter here on Earth is, as far as we are aware, the same as matter in the Sun, matter on Betelgeuse, and matter billions of light years away. Here is a universal that sticks. Matter has the same building blocks wherever and whenever you are.

This is also true of light. It happens that the stuff that makes it up is based on different particles – photons – but once again, all light seems to be the same thing, whether it's streaming through the windows of your lounge or passing through space as the relic of the big bang. It might seem that light is far more varied than, say, an atom of hydrogen, because it comes in many different colours, whereas there's only one type of hydrogen atom (if we ignore isotopes). Light's variety consists not only of the colours we can see, but the far bigger range of different colours we can't see: all of radio, microwave, infrared, ultraviolet, X-rays and gamma rays. Each and every 'colour' is just light.

However, the distinction becomes less clear when you consider what is meant when we talk about the colour of light. When we usually think about something being, say, red, like a postbox, what we mean is that when the white light from the Sun hits it, the object absorbs most of the light, then re-emits only certain frequencies – in this case, mostly red. The colour of light itself, though, is different from the colour of an object. We don't see light as a result of it being illuminated like an object is. In fact, you can't see a beam of light at all sideways on – when you get one of those zippy light beams from a laser in a science fiction film it is almost certainly added afterwards. If it is a genuine laser, you can see the beam only if there is smoke in the room (or some other collection of small particles in the air) and light photons are bouncing off these particles to head towards your eye.

When we see the colour of a beam of light, this is an artificial construct that our brains make, as a result of the response of the different light sensors in our eyes to the light photons that arrive there. This, apart from anything else, is why we can see colours that don't exist in the visible light spectrum, like magenta. Think about a rainbow. Where is magenta? Nowhere. Magenta isn't a colour of light, it's what our eyes detect when they receive white light with the green removed.

Given, then, that the colours we see are subjective, we need a better description for the different colours of light, and traditionally this has been managed by referring to the light's wavelength or frequency, using the model of light as a wave. But when thinking about stuff, it can be more useful to consider light as a collection of particles – photons. And in the case of photons, the 'colour' is a measure of the amount of energy each photon has. Unlike a hydrogen atom, inevitably consisting of a proton and an electron, the colour of a photon isn't an absolute characteristic of that particle – it depends entirely on our frame of reference. We can change the colour of light, for instance, by moving the source, or by moving ourselves with respect to the source. If the source moves towards us, the light is shifted towards the blue because the photons have extra energy; if the source moves away, the light becomes more red as energy is reduced.

So when we say that light comes in different colours, all we really mean is that photons can have different energies – but so can the atoms that make up matter. In practice, atoms have a double dose of energy. There's the energy of movement – as quantum particles, atoms are never totally still and are often jiggling around at high speed. And also there's the energy contained within the structure of the atom. For example, an atom can absorb a photon of light. When it does, an electron in the atom will jump to a different level, gaining potential energy (it is making a quantum leap). So now the atom contains more energy overall.

All stuff, then, comes in standardised packets that can have different energies. But those packets seem to be constructed on the same lines wherever we look. It didn't have to be like this. However, this standardisation makes the formation of complex structures more likely, and it is a boon to scientists, because if every bit of stuff were different, it would be pretty much impossible to make any generalised statements about the universe.

It would be extremely convenient if we could stop our survey of stuff with neutrons, protons, electrons and photons. That's the kind of simplicity we can comfortably get our heads around. However, you are probably aware that there is more going on than this. By the 1930s, a number of other particles had been identified. The anti-electron, or positron, for instance. And the neutrino, a particle that was predicted to exist long before it was observed because of small amounts of energy going missing in nuclear reactions. And during the second half of the twentieth century, the pace of discovery of new types of stuff went into overdrive.

When I was studying physics at Cambridge in the late 1970s, it seemed like every week a lecturer would enter the lecture theatre with more than the usual spring in his step (they were all male), and would announce a newly discovered particle. Things, it seemed, were getting out of hand, with all sorts of new stuff being discovered from cosmic rays – the high-energy particles that crash into the atmosphere from outside the solar system – and from increasingly powerful lab-based accelerators.

Something had to give, and a hint of the way it would do so came out of the consideration of symmetry. There are often symmetries in nature. In the previous chapter, when we had empty space alone, the whole universe had, effectively, just one great overarching symmetry. We tend to associate being symmetrical with simply having an identical mirror reflection. Think of what's meant by saying 'she has a very symmetrical face'. But there are plenty of other symmetries available, from rotation to translation (i.e. moving sideways). In a broad sense, symmetry occurs when you make a change to a system and the outcome looks the same as it did before you made the change. In our universe of empty space, there is total symmetry because we can't do *anything* that will change the way things look.

Once we add in stuff, things get more complicated. A single particle, for instance, still has pretty well every symmetry, because we have no point of reference to say how it has moved. But once we get more than one particle, then there are opportunities to break that symmetry. For example, a collection of particles in the form of an object, with another particle to give it a frame of reference, can lose its rotational symmetry. Think of a letterbox-shaped TV screen. Without an image on it, it only has rotational symmetry when it has been turned through 180 degrees. If it is showing an image (as long as that image is not itself symmetrical), the screen loses all rotational symmetry.

So stuff brings in the potential for symmetry, which in its turn can tell scientists about stuff. By comparing the various particles that were being discovered back in the twentieth century, which had different charges and masses and other less obvious properties like 'spin' (which, confusingly, for particles, has nothing to do with rotating), it was found that there were a number of near symmetries – patterns that suggested particles could be grouped together in systematic fashion. Several physicists independently realised what would take a few decades to prove: that if many of the particles weren't truly fundamental but had sub-components, it would be possible to reduce the number of basic particles back down to a smaller number.

The end product of this rationalisation was the current 'standard model' of particle physics – or to be more relevant to our model universe, the standard model of stuff. It was found that many of the massive particles like protons and neutrons, along with most of the new detected particles, could be made up from varied combinations of smaller particles known as quarks, which had an extra type of charge, known as colour, that came in three 'flavours'. As well as various types of quark, there was the electron and its big brothers, the muon and the tau particle, three types of neutrino and various bosons.

The most familiar of the bosons is the photon, which we usually experience as light, but is also the particle associated with the electromagnetic force. The other fundamental forces had new bosons associated with them: gluons, Z and W bosons. And now, for reasons we will discuss later, we also have the Higgs boson. Throw in antiparticles, which are the basic particle with some properties reversed, and you have the modern standard model. It's not perfect – in fact, it is probably wrong at some fundamental level – but it's by far the best model for stuff that we have at the moment.

Let there be light

We'll have plenty more to think about on how the second kind of stuff, light, behaves when we get on to motion, but here it's worth taking a moment to think about what light *is* over and above a collection of photons. Light is a phenomenon that we experience every day. It enables us to see; it carries the energy of the Sun to power the Earth, and does far more, including acting as a carrier for the electromagnetic force that deals with most of our physical interactions. As we have seen, what was once thought of as just the stuff our eyes can detect – the visible spectrum – extends all the way from radio through to gamma rays. All light.

So we know plenty about what light does for us, but it's much harder to pin down what it truly *is*. It's insubstantial. Unlike matter, we can't touch it. We can only detect it using appropriate sensors, whether they are the visible light sensors in our eyes, the infrared sensors in our skin or the whole range of electronic light sensors we use today, from the cameras in our phones to sophisticated devices like the Hubble Space Telescope.

It was pretty much inevitable that light was first associated with fire. Apart from natural cold light sources like fireflies, and the mysterious distant sources in the sky, fire was the main method humanity had to bring light into darkness. And perhaps it was from seeing the flecks of soot produced by burning torches that the idea first came that light came in the form of particles. We use this word 'particle' so often in physics that it feels like a modern scientific term, but the word dates back at least to the fourteenth century, and it's apt that one of the first recorded uses was in John Trevisa's English translation of Bartholomaeus Anglicus' *De Proprietatibus Rerum* (On the Properties of Things) in which he wrote: 'Sparcle is a litil particle of fire.'

By Newton's day, many thought that light was composed of a spray of tiny massless particles, usually then called corpuscles. But there was a growing suspicion, particularly among Newton's continental rivals including Christiaan Huygens, that light was really a wave, acting in a similar way to the ripples on a pond. It was already known that sound travelled as a wave, and some of light's behaviour, like refraction, the way that it changed direction as it moved into a block of glass, for example, suggested an underlying wave motion.

However, there was a reason for Newton's enthusiasm for his 'corpuscular' theory of light. He knew that light managed to cross empty space. Sound could not do this. In a vacuum, sound failed to carry. This had already been demonstrated by sucking air out of a jar containing a bell. The bell was no longer audible – but it could still be seen. Waves, like sound, needed a medium – a material to do the waving. A wave itself is insubstantial (like light, which admittedly was an encouraging point). It is just a sum of a set of regular movements in a material. But what could that material be when light travelled across apparently empty space?

The answer from Newton's rivals was to suggest the existence of a substance, called the luminiferous aether, which was thought to fill all of space. This aether (or 'ether' as the spelling has become) was strange stuff indeed. It was totally undetectable. It offered no resistance to anything with substance that passed through it. Yet at the same time, it appeared to be totally rigid, even though this made it difficult to understand how a wave could pass through it. If the ether had any 'give' in it, then with time the wave's energy should be lost in the sogginess. But light seemed to go on for ever, unaffected by any losses in the material.

There was another problem with the ether. Over time, it became increasingly clear that light was a side-to-side wave, like a ripple sent down a rope, or the waves on top of a pond, rather than a compression wave that squashed in and out like a concertina in the direction of travel, as sound was known to be. This side-to-side nature was demonstrated, for instance, in the way that light could be polarised. In effect, it was possible with special materials, like the naturally occurring crystal Iceland spar, to separate off parts of a light beam where the side-to-side waves were oriented in specific directions, a phenomenon given the name 'polarisation'.

However, one thing was known for certain about side-to-side waves. They can usually only exist on the edge of something. They don't travel through the depths of the water in a pond, for instance, but only on the surface. This is because the material doing the waving has to have somewhere to go in the side-to-side direction to establish the wave motion. This works fine on the edge of a medium, but try it deep inside the substance and the waving part immediately hits all the other material around it and grinds to a halt. Light, though, merrily passed through the middle of the ether without causing a problem. In fact, it was hard to imagine the ether having an edge.

Although the practical issues of how light could travel as a side-to-side wave remained, it was proved in the early nineteenth century with some certainty that light was, indeed, a wave. Experiments showed that two similar beams of light would 'interfere'. If the beams happened to come together at a point where they were waving in opposite directions at a particular time, the two waves would cancel each other out. But should they come together when waving in the same direction, the two waves reinforced each other, producing a stronger result. Exactly this same effect can be seen in waves on a pond. Drop two stones in and when the waves meet there will be points where they reinforce each other and get bigger, and other locations where they cancel each other out and there is hardly any motion at all.

The final nail in the coffin, it seemed, of light being made up of a stream of particles, came when James Clerk Maxwell identified light as a wave of electromagnetism (see page 116). Maxwell made it clear that a self-sustaining interaction between electrical and magnetic waves could only happen at one speed, which turned out to be what was already known as the speed of light. This made light a self-sustaining oscillating interaction between electricity and magnetism.

And yet, within a few decades, Einstein showed that light did indeed behave in some circumstances as if it were a collection of particles. When light, for instance, produced electricity on hitting certain types of metal, its behaviour was simply not possible if light were a wave. As the quantum theory that emerged from the work of Einstein and his contemporaries became more complete, it was possible for quantum physicist Richard Feynman to say in a lecture: 'It is very important to know that light behaves like particles, especially for those of you who have gone to school, where you were probably told about light behaving like waves. I'm telling you the way it *does* behave – like particles.'

The reason Feynman made this emphatic point to the general public was that he was talking about quantum electrodynamics, also known as QED, the topic for which he won the Nobel

Prize. This sophisticated theory is able to explain all the apparent wave behaviour of light as the action of particles that have phase, an in-built characteristic that alters with time, producing wave-like properties. Feynman always preferred to think of light as being like particles, even though he was among those whose work brought in a different way of looking at light that is now pretty well universal among physicists: considering light to be a disturbance in a quantum field.

Field studies

A field is a very useful mathematical concept that at its most basic is nothing more than a description of a property of nature that has a value at different points in space and time. So, for instance, a weather map displays a kind of field that might show the pressure at different places on the map at a particular point in time. A modern animated weather map can also show how that field changes with time. Taking a field-like look at light, an approach that is applicable to all stuff, light becomes a localised change in the values of the electromagnetic field, a change that moves inexorably with time at the speed of light.

As a mathematical construct, a field can behave however you want it to behave. It is simply a collection of values that is identified at every point in time and space that the field covers. Part of the definition of the field is the type of values that are allowed. It can be totally open – allowing for any value whatsoever – or it can be restricted. For example, the field that describes a chess board is a binary field where values can only have one of two options – white or black. Light, like most of the fields used by physicists, is described as a 'quantum field'.

A binary field is a quantum field, but not a particularly

sophisticated one. A quantum field is one where the value at any point in space and time comes in chunks or packets, known as quanta. That value can be anything (or nothing), but it has to change in quanta, rather than smoothly. So a quantum of light – a photon – can in principle be emitted with any value, but it then has a specific size. If we imagine the value of a quantum field of electromagnetism as a photon of energy E passes by, the field will jump straight from 0 to E at each point the photon passes through and back down again. It is granular rather than continuous. Think of the difference between a fraction, which is quantised because it depends on whole numbers, and a decimal, which isn't.

All the early physicists thought of the descriptions they used for light in absolute terms. So Newton thought that light was *actually* made of corpuscles. And his opponents and later physicists thought that light was an *actual* wave. Even now, when being lazy, physicists may well think that light is a disturbance in a quantum field. However, all of these absolutist descriptions are incorrect. Note, for instance, that Feynman did not say that light was made up of particles, he said that it *behaves like* particles.

Ironically for the phenomenon that allows us to see things, we can't directly examine light. All we can ever do is to construct a model of what it is like. A kind of structured analogy. Early scientists may not have explicitly realised that this was what they were doing, because they weren't aware of anything other than the things they could touch and observe. So they said with conviction, 'Light is a stream of particles' or 'Light is a wave'. But what they were really doing was comparing the behaviour of light to these everyday things, modelling light on them. So in reality they should have been saying, like Feynman, 'Light is like …'

A similar but subtly different thing is occurring when a modern scientist uses a model based on a field. Newton and Huygens used models based on observations of physical objects. We can see waves and particles. A modern physicist is more likely to use a model based on mathematics – and that's what the field model involves. When we take this model-building approach, which is all we can do with something like light, we have moved from the absolute description ('Light is a wave') to a relative description making a comparison with a model ('Light is like a wave'). The model provides our conceptual frame of reference. We have to recognise that this relativistic approach is all that is available to us, because all we can really say absolutely is that 'Light is light', which isn't very helpful.

One of the reasons modern physicists sometimes struggle to remember that fields are still models is that fields are so central to their vision of what the universe is fundamentally like. In building our universe, we started with empty space and now have added stuff. Many physicists, though, would not do this. They would take empty space and add fields. There is even a term for the collection of fields that are considered to make up reality – it's known as 'the Bulk'.

It is possible to envisage that all the stuff in the universe, and the forces that make things happen – even the natural laws – are embodied in a collection of fields, which between them produce the phenomena we experience. This is still a model, but this time the comparison is between nature and a set of mathematical rules. What often happens when combining different elements into a complex mathematical model like this is that some aspects don't work together well. We see this, for instance, in the clash between quantum theory, describing the action of quantum particles (or fields) and general relativity (see page 156), describing gravity. The two simply don't fit together and something will need to give.

One of the ways to fix a problem that arises from this kind of clash of model parts is to add another component to fix the model. This makes the model more complex and more Heath Robinson-like in its working, but it keeps the whole thing going. Historically, when it became clear that simple circular orbits for planets all travelling around the Earth would not work as a model, they introduced the concept of epicycles. These were smaller circles, rotating on the bigger circles, which generated a result that was closer to the observed motion.

Similarly, when the big bang model failed to be able to explain the apparent uniformity in the universe (assuming it is uniform – see page 263), the idea of inflation, where the universe suddenly expanded vastly for no obvious reason, was bolted on to bring the model back into line with observed reality. This isn't the only way to fix a model. Often it is better to throw a failing model away and start again from scratch. However, this may mean scientists giving up many decades of work. Despite their frequent portrayal in the movies as unemotional, scientists are only human, so they often cling on to a theory and modify it, rather than take the dangerous plunge into the unknown. And sometimes the modification of a theory is so successful that this seems worthwhile.

Something like this happened with the field description of the universe. The mathematics predicted that, for instance, the W and Z particles (or rather the disturbances in the appropriate fields) that carry one of the forces of nature, the weak nuclear force, should be massless. Similarly, the quarks that make up protons and neutrons should be massless, though oddly this wouldn't make protons and neutrons massless, because most of the larger particles' mass comes from the energy that holds the quarks together. So rather than start again, theoreticians bolted on another field, now known as the Higgs field, whose only role is to act as a kind of universal gunk to increase inertia, providing that mass.

If the Higgs field does exist, it should be possible to detect disturbances in it – which can be observed as particles. And this is why we got the search for the Higgs boson. Of itself, the boson isn't something that has an obvious role like, say, an electron or a photon. It's not the Higgs boson that produces the missing mass. But it would be expected to exist if the Higgs field were there.

It might seem after the Large Hadron Collider results in 2013 that there can be no doubt about the Higgs field, but detecting a particle that matches expectations for what a Higgs would be like does not somehow make the field model an absolute reality, any more than detecting a photon or a light wave does it for the electromagnetic field. It also ought to be stressed that, despite the way it was often reported, the experiment did not show that there was a 1 in 3.5 million chance that the Higgs did not exist – it showed there was a 1 in 3.5 million chance of that result occurring purely randomly without a particle to cause it. There is no way to prove it was a Higgs, merely that it fits what would be expected.

This is not being critical of quantum field theory. It is the best current model to describe the key behaviour of stuff (though in some circumstances waves or particles can be a lot easier to use). And with the Higgs field added, it is pretty close to a good mathematical description of what is observed, if somewhat messy. But we shouldn't confuse it with an actual description of what stuff *is*, any more than we confuse an address with the house that the address identifies.

May the force be with you

Several times now we have come across the idea of fundamental forces of nature. From the field viewpoint, each is a field that fills the universe. But when we look from the viewpoint of adding stuff to our empty space, the four forces – gravity, electromagnetism, the strong and weak nuclear forces – are inherent aspects of stuff that describe how it behaves. (We will need to add both time and motion in the following chapters to see much of this behaviour.)

Of these four forces, gravity is significantly different from the others and provides a natural tendency for stuff to be attracted to other stuff. We will cover this separately in Chapter 6. Next most familiar is electromagnetism. This enables some kinds of matter to attract or repel other kinds, and accounts for the interaction of light and matter.

The remaining two forces are nuclear forces, typically functioning in the incredibly close confines of atomic nuclei. The strong nuclear force provides the glue that holds quarks together in particles like protons and neutrons, and that holds the nuclei of atoms together, despite the electromagnetic repulsion of positively charged protons.

The weak nuclear force does not obviously provide the usual attraction or repulsion, but rather is responsible for the switching of some particles of stuff into other related particles. This role means that the weak force is involved in the kind of nuclear decay that fuels nuclear power stations and atomic bombs.

The combination of the four fundamental forces and the particles of stuff come together to provide a mostly complete description of the behaviour of the universe. However, in practice, there are usually many particles present, and the behaviour of quantum particles is dependent on probabilities rather than absolute values, making it impossible to treat real-life examples using such detailed considerations. Approximations and simplifications become necessary.

Other matters

Whichever way we decide to describe stuff, matter comes in two, or possibly three, or maybe even four broad types. The two that we are absolutely certain about are matter and antimatter. Back in the late 1920s, theoretical physicist Paul Dirac, probably the least known of the great contributors to quantum theory, was attempting the messy process of combining quantum theory, describing the behaviour of particles like the electron, and the special theory of relativity – an essential addition, as electrons and their like often move sufficiently quickly that this modification of Newton's laws comes into play.

Dirac struggled for some time to come up with an appropriate equation, but finally produced his masterpiece, which succeeded in describing the behaviour of electrons and similar particles, even if they were moving at near the speed of light. But the equation came with a painful price tag. To be able to describe the electron properly, the particles had to be able to have either positive or negative energy. However, this proved challenging, as it seemed to suggest that electrons should be able to lose more and more energy as they plummeted down to negative values – effectively each electron would become an infinite energy source, which clearly didn't happen.

Dirac's solution to the problem was radical. He imagined that the universe started off with an infinite 'sea' of electrons, filling all the negative energy positions available. This meant that an electron could never drop below having zero energy – there was nowhere for it to go because all the spaces were already filled by the sea of pre-existing electrons.

This is, to say the least, a bizarre and hard-to-accept solution, although it did make his equation usable. And with the negative energy sea in place, the equation made impressive predictions about the behaviour of the electron that matched well with what was observed. However, Dirac went further. If there were such a sea of electrons filling the negative energy slots, then it made specific, testable predictions. If an electron is given energy by, for instance, an incoming photon of light, it jumps up to a higher energy level. If one of the negative energy electrons in the sea were hit by an appropriate photon it should jump out of the sea, leaving a hole behind it.

Dirac gave some thought to how such a hole would appear. It was an absence of a negatively charged, negative energy particle. He realised that this would behave identically to the presence of a positively charged, positive energy particle. By zapping a negative energy electron with energy, the result seemed to be the production of an unheard-of particle – the same as an ordinary electron, but with a positive charge. This would be an anti-electron, or to give it the name that it soon received, a positron.

What's more, an ordinary positive energy electron* could drop into the hole, giving off energy and making both the electron and the hole disappear in the process. This would be the equivalent of the electron and the positron – respectively matter

---

* It's important not to confuse electrical charge and energy. An ordinary electron has a negative electrical charge, but it has positive energy: the energy represented by its mass, the kinetic energy of its movement and its potential energy if it is part of an atom.

and antimatter – coming together and annihilating to give off energy.

Dirac originally thought that these positive particles were the protons found in atoms, but as anti-electrons were effectively a missing negative energy electron there was no good reason why they should not have the same mass as an electron, and if that were the case, his model suggested that electrons and protons should annihilate each other, rather than make stable atoms. With a twist of fate, Dirac, who spent most of his working life in Cambridge, was on a sabbatical at Princeton when American physicist Robert Millikan came to Cambridge to present work that would reinforce Dirac's idea. Millikan's student Carl Anderson was studying the effects of cosmic rays, high-energy particles that stream towards the Earth from deep space.

Anderson had set up a cloud chamber, a device that produced trails of droplets when particles passed through it. He discovered a number of times when electrons formed, they were paired with another particle which curved off in the opposite direction under the influence of a magnetic field, showing that it had the opposite electrical charge. With increasing evidence from better-quality equipment it was obvious by 1933 that a positively charged equivalent of the electron had been discovered, produced in a pair with an electron, a particle that seemed a perfect match for Dirac's anti-electron.

Although Dirac's original prediction was based solely on the electron, and formulations of his theory would soon be made that did away with the need for the negative energy sea, it was realised that there was no reason why there could not be antimatter equivalents of all the fundamental particles. This is fairly obvious with, say, a proton, where an anti-proton is its negatively charged partner – and at CERN, antimatter atoms are regularly produced from a combination of positrons and anti-protons.

What is less obvious is that there are also, for instance, anti-neutrons. Even though there is no negative charge to reverse, the anti-neutron has reversed values of many of the neutron's other characteristics. It has even been speculated that anti-atoms could have an inverted reaction to gravity, being repelled gravitationally by ordinary matter, though as yet insufficient antimatter has been made to test this out, and most physicists consider the outcome highly unlikely.

The existence of antimatter has not just been proved experimentally, but also turned out to be highly useful when building a model of what happened to stuff in the early times of the universe. The big bang model required the whole universe to originate from pretty well a point source. Clearly it isn't practical to cram all the atoms now in the universe in one place. However, the theory required that in those early times there was only energy, which had no space requirements.

Just as a particle and an antiparticle could combine to produce energy, so energy can transform into a particle/antiparticle pair. The existence of antimatter provided a mechanism for matter to come into being, if there was already sufficient energy. But there was one significant problem. There ought to be an equal quantity of antimatter as matter in the universe, ready to recombine and destroy all the matter. Thankfully for us matter-based life forms, in reality there is far more matter than antimatter out there. So what happened to the rest of the antimatter?

As is usually the case when looking back to the early days of the universe, theories are highly speculative, but a couple of possibilities are that the matter and the antimatter became partitioned somehow – so the antimatter *is* out there somewhere but out of reach – or that there was some slight lack of symmetry in the process that meant there was a tiny percentage of extra matter available and this tiny percentage is what we now see as making up the entire contents of the universe.

Going dark

Antimatter is still conventional stuff, still subject to the same forces as ordinary matter. But the other possible types of matter are far more exotic, at least when viewed from the relative position of conventional, matter-based organisms like us. These other types of matter are dark matter and anti-dark matter. As we have seen, dark matter is a hypothetical 'other' kind of matter that only interacts with the matter from which stars and planets and people are made via gravity.

As dark matter doesn't react to either the electromagnetic or strong nuclear force, the room in front of you could be full of dark matter now, flying without any noticeable effect through your body, and you would not be conscious of it. This is not as unlikely as it sounds. We do know that there is a type of conventional matter in the form of particles called neutrinos that have only a tiny ability to interact with atoms. Billions of neutrinos pass through your body from the Sun every second, yet you are unaware of them.

However, even the Sun's torrential output of neutrinos is negligible when set alongside the quantity of dark matter that is assumed to be in the universe. Dark matter makes itself known by the gravitational effects of large clumps of it, particularly on the scale of galaxies. There, its presence is felt so strongly that there is estimated to be around five times as much, by mass, of dark matter in the universe as there is conventional matter.

This ratio is often stated, but just take a moment to consider the implications of those numbers. It appears that even though to get the 'conventional matter' total we are adding up every star, planet, black hole, speck of dust and gas in the universe, we come up with only around one fifth of the amount of stuff that is out there in the form of dark matter. This isn't a minor tweak to our understanding of reality. It blows a universe-sized hole in what we actually know about. Despite those billions of galaxies, each containing billions of stars, planets and (at least on Earth) living things, the whole lot is dwarfed by the quantity of dark matter in the universe.

It would be rather useful, then, to know what dark matter is. But there's a problem. How do you really get to understand something you can't see, or touch or interact with apart from indirectly observing its gravitational impact when in very large clumps indeed? In a sense, that name 'dark matter', impressively endowed with the feel of broody horror fantasy, is a bizarre misnomer. Dark matter is about as un-dark as you can get – it is totally, entirely invisible.

However, particle physicists and cosmologists are always making observations that are necessarily indirect. We can't see or feel an electron or interact in any direct way with a black hole. Yet it hasn't stopped a huge amount of work being done on them. And though it is inevitably speculative, there has been some serious thought put into the nature of dark matter. To see where these ideas come from, we need first to look for the evidence that dark matter exists at all.

It might seem that dark matter is a new obsession of cosmologists – no one taught us about it at school, after all. But the concept of dark matter (or, to be precise, *dunkle Materie*) goes back to the 1930s, when the Swiss astronomer Fritz

Zwicky, working at the California Institute of Technology, realised that there was something very odd about a group of galaxies called the Coma Cluster. It seemed impossible for it to stay together. Just as a piece of clay on a potter's wheel will only stay together if the cohesive stickiness of the material is stronger than the desire of bits of clay to carry moving in a straight line and fly off the wheel, so a galaxy or a cluster of galaxies like Coma is in a balance between the tendency of the stars or galaxies to fly apart and gravity's ability to hold them together.

Zwicky could not find enough stuff in the cluster to keep it in one piece – it needed far more mass to create that gravitational attraction than any sensible approximation to the mass of visible matter in the stars that made up the galaxies. Of itself, this does not require our modern concept of dark matter as a new and different kind of stuff that only interacts with ordinary matter by gravity. After all, not everything in a galaxy is visible from a great distance. There are planets, for instance, which are far too small to detect in galaxies outside our own, although their contribution to mass is quite small.

Then there's dust and gas. We're used to these being true lightweights. But remember that stars and solar systems are formed by the coalescence of dust and gas. There's a whole lot of it out there. And, of course, there is truly dark stuff, if not dark matter in the modern sense, in black holes. Zwicky wasn't aware of it, but we now know that a galaxy typically has a vast black hole at its centre with the mass of millions of stars. However, Zwicky's estimate for the mass required was so great that, even allowing for all this, it seemed hard to reconcile the amount of stuff involved with any sensible assessment of what makes up a cluster of galaxies.

Zwicky was something of a loner who had a tendency to throw out a mix of brilliant and wacky ideas, meaning that it was easy for his observations to be practically ignored, and it wasn't until the 1970s that American astronomer Vera Rubin at the Carnegie Institute made a similarly puzzling observation of the way that stars orbit around in spiral galaxies. Although they vary in structure, galaxies (like solar systems) tend to be roughly disc-shaped because of the way that they are formed, and often have spiral arms with a bulge near the middle.

Like pretty well everything in the universe (see page 114), galaxies have a tendency to rotate, and when a disc rotates we expect to see varying velocity, depending on how close to the centre a part of the disc is. Just think of a vinyl record. If it's rotating at, say, 45 revs per minute, then in just over a second, the outside edge of the disc has to travel the circumference of the disc. But the edge of the label in the middle has much less far to travel in the same time – just the circumference of the label – so it has to be moving a lot slower.

I couldn't use the now more familiar CD or DVD as an example, as they change rate of rotation depending on where the read head is, to keep a roughly constant velocity at that point. But it turned out that galaxies were more like vinyl discs than CDs, in that when Rubin and her colleagues looked at a galaxy they found that stars had roughly the same rotational rate near the outside of the galaxy as they had near the centre. This meant that the outer stars had to have a much higher velocity than otherwise expected, to get around the whole of the perimeter of the galaxy in the time available for the rotation. They were moving so fast that they shouldn't stay in place.

This is the kind of anomaly that scientists love. You don't win fame and Nobel Prizes by showing that what everyone thinks happens is true – it's for discovering the unexpected. And this was the unexpected in spades. Rubin's team worked out that there needed to be around five times as much undetectable matter in a galaxy as there was ordinary material in order to keep it together and able to rotate in this uniform fashion.

A range of other large-scale astronomical observations appeared to confirm that there is just far too much mass out there to be accounted for with the traditional matter content of a galaxy. A good example of this is in gravitational lensing (another idea of Zwicky's). As we will discover in Chapter 6, Einstein's work on gravity made it clear that objects with mass warp space and time. This means that when light from a very distant object passes a closer object, it is bent in towards that object. This is not unlike the way that a lens bends light, and with enough matter, the object can act as a lens, focusing the image of a distant object behind it.

Depending on the exact configuration, this gravitational lensing effect can produce multiple images of the distant object, making it obvious that the effect is occurring – and making it possible to deduce the mass of whatever is acting as a lens. And if that lens happens to be, say, a galaxy, it is all too obvious that it has more mass than any sensible combination of the number of stars and other components in it is likely to contain.

Totally cosmic background

Although observing galaxies and clusters can give a feel for the amount of dark matter present in them, it's not enough to get a picture of the universe as a whole – because there's a lot of universe that isn't occupied by a galaxy. Luckily, cosmologists have a way to get a sense of the overall quantity of dark matter present, assuming it has been around as long as ordinary matter, from the cosmic microwave background radiation.

This is the microwave-wavelength light that permeates the universe – the glow that started out when matter first formed atoms and the universe became transparent, around 380,000 years after the big bang. Before then, the universe was full of plasma, like a star, and was anything but transparent. The radiation is almost the same whatever direction you look in. That's how it was identified as what it is, rather than light from a specific source. But it has tiny variations, which have proved a rich source of information about the early universe.

Originally detected from a ground-based radio telescope, the radiation has now been mapped with some accuracy by a series of satellites. The way the egg-like map of all directions is presented is a little misleading. It looks like there's a huge variation in the radiation, but the most extreme variations reflect a difference of only around 1 in 10,000. Although these maps are presented as clear and definite, there is always a danger that some of the apparent patterns are artefacts caused by the significant processing required to remove noise and the influence of dust in space, but the indications are good that the maps give a reasonable indication of the state of the universe when it was just one third of a million years old.

The variations on the map indicate differences in composition of the early universe, very subtle distinctions, but providing variation that would become the seeds of the early galaxies. And by using indirect measures, it is possible to estimate the amounts of dark matter to ordinary matter present, coming up with a ratio of around 26:5 – once again, more than five times as much dark matter as the ordinary stuff.

One significance of this abundance of dark stuff is that

without all that dark matter it's entirely possible that the early components of the universe would have been unable to form stars and galaxies until much later, as ordinary stuff alone would not have had enough gravitational attraction to overcome the natural tendencies of the high-energy atoms to fly away from each other, buffeted by the stream of radiation they emitted. Not only was there a lot more dark matter than ordinary matter to help the galaxies form, but it was immune to this battering from light as it isn't affected by electromagnetic radiation.

Candidates for darkness

As mentioned in Chapter 2, it's just possible that dark matter doesn't exist and all the effects we attribute to it are produced by a variation in gravitational action on the scale of galaxies and clusters. But assuming, as most astronomers do, that it does exist, we then have to consider just what it is that we are dealing with. To which the simple answer is 'we don't know'. We have never seen or directly interacted with any dark matter. Not one particle. There are plenty of experiments under way attempting to do so, as, despite its reluctance to interact, it is just possible that small interactions with ordinary matter would be possible. But all attempts have so far turned up nothing. We know a fair amount about how dark matter behaves from its gravitational interactions on a large scale, but no certainty about what it is. However, there are, inevitably, theories.

The most strongly supported theory for a good few years now has been that dark matter is made up of WIMPs. (That's 'weakly interacting massive particles'.) We know a lot about the main particles that make up ordinary stuff and that form our standard model of particle physics. But WIMPs are something else. They are neither bosons, like photons or the Higgs boson, nor fermions like electrons and quarks. They are something else entirely that doesn't have a place in the current standard model. If WIMPs were shown to exist for sure, we would need either to accept that they are just outside the model or to build a totally new one.

There was some attempt initially to fit WIMPs into the standard model, because that would keep things simple. For example, it was suggested that what were thought to be WIMPs might actually be neutrinos. These are particles of ordinary stuff that we understand well. And as neutrinos have no electrical charge and very little interaction with ordinary matter, they seem reasonable dark matter candidates.

There was initially one snag. For a long time it was thought that neutrinos were massless. Given that the main point of dark matter is to add extra mass to the universe to account for the behaviour of galaxies and galactic clusters, a massless particle wouldn't have been much of a candidate. But it turns out that neutrinos do have a tiny mass. Unfortunately, however, this mass doesn't seem to be anywhere near enough to account for the gravitational attraction of dark matter – and neutrinos do have *some* detectable interaction with ordinary matter, pretty much ruling them out.

What's more, neutrinos tend to move at extremely high speeds – but for dark matter to coalesce and pull together galaxies it would have to be moving relatively slowly. Hence cosmologists tend to refer to 'cold dark matter'. 'Cold' here is being used to mean slow-moving. Temperature is a measure of the energy of movement and excitation in a material, so in cold matter, the particles will be relatively slow.

A half-in, half-out option to try to keep dark matter vaguely associated with the standard model, and which still has popularity with some physicists, is the neutralino. This is one of a whole host of extra particles that would theoretically be added to the standard model if an approach called supersymmetry applied, which requires every particle to have a 'supersymmetric' partner. The neutralino would be a good candidate as it is a massive neutral particle that would have limited interaction with ordinary matter – but despite years of efforts to find evidence, no experiment has ever detected a supersymmetric particle and the theory is generally considered to be in decline.

Another model has been based on ultra-light but hugely populous particles called axions (though most attempts to make use of these struggle because they are difficult to match to the observed behaviour of dark matter), while there are even attempts to find a way to use basic ordinary matter to explain at least some of dark matter's apparent capabilities. This last is given the deliberately WIMP-opposing name of MACHOs, the clumsy acronym for massive compact halo objects. Unfortunately the models of how matter formed in the early universe do not allow for the production of so much ordinary stuff, bearing in mind the 26:5 ratio. There could always have been other processes, but this would mean a huge upheaval in the models of the early universe just to make MACHOs possible.

Although MACHOs have been pretty much ruled out, however, they are by no means the only alternative to WIMPs, axions and the like. And this becomes clear when we realise just what strong and rather strange assumptions were made when coming up with an idea like the WIMP. To keep things simple, physicists were looking for a single type of particle that could not interact with ordinary matter or itself other than as a result of gravity. It certainly couldn't have the ordinary interactions we are familiar with between matter particles, because these depend on electromagnetism, which we know that dark matter appears to blithely ignore. But we have to be aware that this concept of dark matter was based purely on an idea of 'keeping things simple'. And all our experience with ordinary matter is that things aren't particularly simple. So why should we expect the far bigger population of dark matter to be any different?

Specifically, the standard model for ordinary stuff contains seventeen different particles and to greater or lesser extents those particles are influenced by the four different fundamental forces. Yet in dreaming up dark matter, it has been largely assumed that there is only a single type of dark matter particle, feeling the influence of just one force – gravity. Frankly, this displays an uninspiring lack of vision. Imagine for a moment you existed in a parallel universe where interactions with normal matter were as weak as ours with dark matter. You would be very wrong if you assumed that the stuff in our universe was caused by a single 'ordinaryon' particle with no self-interactions. And there is an increasing amount of work being done looking at the possibility that dark matter, like ordinary stuff, has a more complex set of building blocks.

The implications are quite extraordinary. It's like a curtain being drawn aside, a sudden reveal of impressive proportions that makes a huge amount of sense. Physicist Lisa Randall has described those who insist that ordinary matter should be considered the dominant stuff with dark matter relegated to being a one-trick sideshow as 'ordinary matter chauvinists'. With our seventeen stuff particles, why did we ever think that there was just one dark matter particle, apart from keeping life simple for ourselves? Of course, there is no reason why dark matter's set of particles should parallel those of the ordinary world. There could be fewer or more. But it's entirely possible that dark matter is not a collection of uniform material.

For that matter, if you will pardon the pun, why assume dark stuff is just matter? We have accepted that we need to think about wider categories of stuff in the ordinary, non-dark world. Why should the same not apply to dark stuff? For instance, just as we have antimatter in our universe, it's entirely possible that there is anti-dark matter too. And things would get even more interesting if there were a dark equivalent of, for instance, light. The concept of dark light takes a little getting your head around, but if we call it instead 'dark radiation' it becomes less confusing. Just as ordinary matter can give off electromagnetic radiation, dark matter could give off its own form of radiation, undetectable with our ordinary matter instruments.

In the ordinary world, electromagnetic radiation is not just light, but the carrier of the electromagnetic force. And in the dark world, whether carried by dark radiation or some other means, again there is no reason that there couldn't be dark forces – either paralleling those in the ordinary world or quite different. These forces would mean that some dark matter particles could interact with each other in ways other than by gravitation, something not allowed in the simple single-particle models. It's likely that this wouldn't be true of all dark matter particles, as such a level of interaction would probably be detected even with our very indirect means. But again, not every particle in our ordinary stuff world responds to all four forces – and the same would very likely be true in the dark universe.

Take this picture to the extreme and there is no reason in principle why there couldn't be dark planets orbiting dark stars (which, of course, in the dark universe are not 'dark' at all, but pouring out dark radiation). It's even possible to go the whole hog and speculate that there could be dark life in that dark universe. Science fiction has often played with the idea that there could be such a thing as a 'parallel dimension' where whole different worlds exist alongside our own universe and yet are unable to be in contact with it. A dark universe with dark life on dark planets would provide a real basis for such a second universe, intermingled with ours and yet not interacting, without the need for spurious parallel dimensions.

Is this speculation gone wild? Probably. There is no evidence at all for any of these features of dark matter. And as we'll discover in Chapter 7, life is very difficult to get started in our universe, and this may well be true even if there were dark planets. However, the possibility of a dark universe that is made up of more than a single type of particle is far more than science fiction; it is a possibility supported by the evidence, rather than running contrary to everything we have ever observed. It appears that dark matter is out there. And that could be all there is to it. But there is some intriguing evidence that dark matter may have a degree of self-interaction. And the mere possibility that there could be far more is surely one of the most intriguing aspects of dark matter, which at the time of writing is surprisingly infrequently mentioned.

The mass effect

So now we have a picture of all of stuff – matter and dark matter, antimatter and possibly anti-dark matter. Plus light, and the other aspects of stuff that we can't truly describe as matter. There are two essentials to stuff – what it's made of, and how it behaves. I have already frequently made use of one of the key measures of the behaviour of matter, its mass, without explaining what was being described. This is dangerous, as mass is a concept that often gets confused with weight – a confusion that isn't helped as we typically use the unit of mass for weight, even though that unit isn't strictly applicable. Mass tells us how much of a particular type of stuff is present and dictates how it will behave in two circumstances, described as inertial and gravitational, which we'll come to in a moment. Weight, by comparison, is the force generated by a mass when it is in a specific gravitational field.

So, for instance, my mass is around 80 kilograms. It would be the same if I were on the Moon or floating in space. Strictly speaking, on Earth I weigh around 785 newtons, which is the gravitational force I feel pulling me towards the Earth. However, this is calculated by multiplying my mass by the gravitational acceleration on the surface, which is around 9.81 metres per second per second (m/s^2). In other words, every second something is falling it goes 9.81 metres per second faster. As the acceleration is pretty much the same wherever you are on the Earth (it varies with altitude, but only by a small amount), we tend to miss that bit off and say that I *weigh* 80 kilograms. But this is misleading, as weight is entirely dependent on the gravitational frame of reference.

Were I to travel to the Moon, because the amount of stuff in me does not change, my mass, as we have seen, remains the same. But the acceleration due to gravity on the lunar surface is about 1.6 m/s^2 – around one sixth of that on the Earth's surface. So up on the Moon I would weigh around 128 newtons. As we cheat and divide weight by the acceleration on the Earth's surface, we tend to say that on the Moon I would weigh 13 kilograms. When Newton was working on the *Principia*, his masterwork on mechanics and gravity, weight was the only measure in common usage. Yet, surprisingly, even with the technology of the day, Newton did know that an object's weight could vary.

Astronomers like Newton's contemporary Edmond Halley were used to travelling to different locations around the world to make observations of the night sky, and when they did so, they discovered that pendulums (becoming common following the invention of the pendulum clock in 1656 by Christiaan Huygens) moved at a slightly different rate, showing that the pendulum weight was varying depending on location. Newton needed a measure of the quantity of stuff that did not vary from place to place for his work on gravitation, and so he devised what was then a new concept – mass. In the *Principia* he says:

> Quantity of matter is a measure of matter that arises from its density and volume jointly … I mean this quantity whenever I use the term 'body' or 'mass' in the following pages. It can always be known from a body's weight, for – by making very accurate experiments with pendulums – I have found it to be proportional to the weight, as will be shown below.

As alluded to above, an object's mass comes into play in two separate circumstances where the mass could, in principle, have two separate values – one mass that determines how an object behaves under the force of gravity and a second that is its inertial mass. This would be the one that determines how much force it takes to get the body moving with a particular acceleration. As it happens, these two masses have identical values, so we can conveniently just talk of a body's mass without worrying which we are referring to. But in principle they could have been different, which would have made physics significantly more complicated.

Fixing the unit

Compared to specifying the speed of light (see page 27), having a specific unit for mass (or initially of weight in a particular location) is a surprisingly fiddly business. Traditionally, the approach taken was to have a master object that defined that mass, which for the metric system that is the standard for science was a cylindrical chunk of platinum–iridium alloy that was used to define a mass of 1 kilogram and against which, in theory, everything else was compared. Forty of these platinum–iridium kilograms were manufactured in France in 1879.

This kind of standard is fine on a local basis, but makes it difficult to have a true and uniform standard that applies around the world. So, for instance, at the time of writing there are still very small differences in the reference mass objects used by France and Australia. It has been pointed out that a kilo of French cheese is around a millionth of a gram lighter than a kilo of Australian Vegemite.

The platinum–iridium kilograms provide a perfect illustration of the difficulties of basing a unit of measurement on an actual object. Every time the cylinder is handled there is a fear that it could lose a tiny quantity of its surface, which would reduce its mass; or, just as bad, even the relatively unreactive alloy could get some kind of deposit from the air, increasing the mass. What's more, the official cleaning regime has hit problems, as it was originally specified that the cylinders should be cleaned with chamois leather. Unfortunately, genuine chamois leather is no longer legally available, as the goat the leather used to be taken from has become an endangered species.

Work is under way to replace all units that are currently based on a physical object with ones derived from a specific and invariant natural quantity, such as the mass of (say) a specific number of neutrons. But as yet we are still reliant on those lumps of metal. The best-known approach being used to try to replace the cylinders is the so-called International Avogadro Project. This uses a pair of 1 kilo spheres made from the stable silicon-28 isotope. Because the spheres consist of single isotopes and have a highly uniform crystal structure, it's possible to calculate the number of atoms in the sphere, and hence to have a link between the kilogram and a fundamental natural quantity.

However, while the approach is simple in concept, ensuring the perfection of the silicon spheres is anything but easy, and the idea has been overtaken as a future standard by an approach known as a watt balance. This balances a weight, under the influence of gravity, against the force generated by an electromagnet. With some fancy quantum equipment to precisely measure the electrical energy involved (the balances end up using two obscure quantum phenomena, the Josephson effect and the quantum Hall effect), the mass can be established as a relationship involving a universal constant that is widely used and measured.

This value is the Planck constant, more commonly used to provide, for instance, the ratio of the energy of a photon to its frequency. Using a watt balance is not exactly a direct and obvious way to define the kilogram, but it seems to be one that holds out the best hope for a clear measurement linked to a fundamental of the universe, and is expected to be in place as the new standard by 2018.

Heavy light

As for other types of stuff, like light, it might seem that mass is an irrelevancy, because photons don't have a mass. This appears at first to be an obvious assertion, as it's hard to imagine that, say, a box full of light would be harder to move than an empty box. But it's easy to be fooled when dealing with extremely small values, so we need to have more than a feel for this.

As we will discover when we get on to Chapter 5, movement has an impact on mass. Mass is not an absolute, but something that can vary in a relativistic fashion. Were it possible to get hold of a static photon, then mass would be a meaningless concept for it – in this sense it genuinely is massless. But in practice photons are always moving – and for practical purposes this movement means that they act as if they had mass.

One way of looking at this is through the equation we have already met, $E = mc^2$. Photons of light have energy – they are, in effect, a form of pure energy. The energy we receive from the Sun that keeps Earth alive arrives in the form of light. And that equation tells us that the tiny packet of energy in a photon of light is equivalent to an even tinier (because we have to divide it by the speed of light squared) amount of mass.

Despite being nominally without mass, photons are influenced by gravity. As we saw with gravitational lensing, when a beam of light passes close to a massive body, its path is bent away from a straight line, just as is the path of an orbiting satellite (though on a much smaller scale). Harder to get your head around, light also falls due to gravity as it goes on its way on the surface of the Earth.

Consider a little thought experiment. We are going to compare three things. We will fire a gun horizontally. At the same moment, we will drop a bullet from exactly the same height as the gun, and shoot a beam of laser light horizontally. If the Earth were perfectly flat, all three would hit the ground at exactly the same time. In practice, light goes so fast that it will be well away from the curved surface of the actual Earth before it has a chance to reach the ground. But it does fall at the same rate as the bullets.

Similarly, a photon of light has the kind of mass effect we experience as inertia. We can see this in the form of a property called momentum. With a normal piece of matter, the momentum is the mass times the velocity. This measures how much 'oomph' a moving body has. And light has momentum too, related to its frequency or wavelength when seen as a wave, or the energy of the photons.

If you have a flow of bodies with momentum hitting an object, this results in pressure, which means that light can produce pressure, just like a flow of gas molecules (though it is much weaker). This is the idea behind solar sails, which are huge expanses of material used in space to pick up light pressure from the Sun and use it as motive power. It used to be thought that the toy called a Crookes radiometer demonstrated light pressure on the desktop. A radiometer looks a bit like an old-fashioned light bulb – a glass bulb with most of the air pumped out, and in the middle a wheel with paddles that rotates when exposed to light.

The paddles are black on one side and white on the other. The idea was that the black paddle sides absorb the light, but the white sides reflect the light. This means there should be a net pressure on the white side, starting the paddles to rotate away from that side. Unfortunately, they go the other way round. The actual effect is because the black sides, absorbing light, warm up. As a result they warm the air in contact with them – because the bulb hasn't got a complete vacuum in it. This means more collisions with air molecules, so the black sides get more of a push, and the paddles rotate. The light pressure is just not strong enough to produce motion.

Electromagnetic action

Almost all our interactions with matter and light involve electromagnetism. As one of the four fundamental forces of nature alongside gravity and the strong and weak nuclear forces (which come into play at the level of the components of an atom), electromagnetism is an essential component of our model universe. Although we tend to spot the influence of gravity more explicitly as an active force, far more of what we experience day to day comes down to electromagnetism. Interacting with stuff is an electromagnetic business.

This is pretty obvious when using electricity or a magnet, but it equally applies when, for instance, we pick up an apple or sit on a chair. Of themselves, 'solid' objects are almost entirely empty space. If an atom were blown up to the size of a large building, the 'solid' nucleus would be about the size of a pea. The rest of the atom, apart from one or more electrons in some kind of fuzzy existence around the outside, is empty space. And solid objects have far more space between atoms than within them. So when you attempt to sit on a chair, the most obvious outcome should be to slip straight through it.

In reality of course you don't do this – and the reason a chair can support you is electromagnetism. It is the repulsion between the positively charged nuclei of all the atoms involved that prevents them getting close enough for you to pass through. You float fractionally above a chair on the reluctance of the similar charges to come together. And this kind of electromagnetic interplay is responsible for almost all interactions with stuff.

The processes are more subtle when light interacts with matter. But we're still dealing with an interaction as a result of their electromagnetic natures. And the electromagnetic contribution to matter is not all about repulsion. Solid objects are held together by the attractive side of electromagnetism. This effect is weaker in liquids, but still there, enabling the liquid to stay together – and providing interesting extra features like hydrogen bonding between the molecules of water, which pushes up its boiling point. Without this, water would boil below room temperature and we wouldn't have liquid water on Earth.*

◇◇◇◇◇

Stuff, then, is the essential of everyday life, yet under the surface it is far more interesting and complex than it first appears. You might think that you are looking at a lump of cheese or a piece of wood or a shaft of light, but the nature of stuff and how it plays its part in our overall universe is surprisingly complex. When we add stuff to our universe we need to add both its component parts and the forces that govern its complex interactions.

---

* Electromagnetism is also responsible for other attributes of stuff. When, for instance, we bend a metal spring and it pulls back into shape, this elasticity is a result of the attraction between atoms pulling them back together. Funnily enough, given their name, the best-known example where this electromagnetic elasticity *isn't* in action is in elastic bands. In these, the rubber is made up of molecules in long chains that are naturally full of kinks.

When we stretch a rubber band, the kinks in the molecules are partially straightened out. Now, despite a solid like rubber appearing to be static when seen with the naked eye, if we could zoom in and take a look at the molecules that make it up, they are constantly jiggling about. Which means that the nearby molecules keep bashing into a stretched molecule and push the kinks back into it. This shortens the molecule. So the rubber is always trying to fight against the stretch, not because of electromagnetic attraction between molecules, but because the heat of the room means that the molecules are jiggling about and colliding.

However, by bringing in those interactions we are getting ahead of ourselves. As yet our model universe contains only space and matter, and that is not enough for things to be able to happen. Any interaction will usually involve change. If we unpack that word, we expect something about the universe to be different, not in two spatial locations, but at points that we are comparing in a totally different dimension. Change requires that we add time into our model.

That is not going to be an easy step to take. If we don't think about it deeply, time seems to be an ordinary part of everyday experience. Yet when we try to examine it in detail, when we have to be specific about what time is and how it works, this next component will prove far harder to add to our universe than was stuff.

4 Time

◇◇

By introducing stuff to our toy universe we have made space more accessible. We can give our universe scale and we have the ability to form frames of reference to put position into context. We can imagine structures and objects. But space and stuff alone are not enough to construct a working universe. The very concept of working or functioning implies that there is the opportunity for change. And if we want our universe to be capable of change, rather than being frozen and immutable, it seems that we also need time. I say 'seems' because some physicists argue that time does not truly exist. We will come back to why they say this, but first we need to be sure what we mean by time.

Without matter, our empty space could manage comfortably without the concept of time as there was nothing that could undergo change, nothing with which to mark any difference in time. Time, in fact, would have been meaningless, unless the universe came into being or ceased to exist at particular instants. But matter, and its gift to space of relativity, also adds value to time. Unless all matter is simply there, unchanging and effectively worthless, there is a need for time – for instance, to establish when a particle appears. Time and change go hand-in-hand if there is to be any context. Yet unlike matter, time is a slippery, diffuse concept. It's something we are constantly aware of, yet any attempt to pin down the nature of time runs up against serious difficulties in our ability to model it, let alone describe it.

St Augustine of Hippo, a fourth-century bishop and one of the best thinkers of the early Christian church, made the very apt comment:

> What is time? Who can explain this easily and briefly? Who can comprehend this even in thought so as to articulate the answer in words? Yet what do we speak of, in our familiar everyday conversation, more than of time? We surely know what we mean when we speak of it. We also know what is meant when we hear someone else talking about it. What, then, is time? Provided that no one asks me, I know. If I want to explain it to an inquirer, I do not know.

We tend to think of an obsession with time as being a modern thing. The assumption is that it was only with the advent of clocks and a time-driven society of clock-watchers that the interest in time arose. Before then, it seems natural to us that a pre-industrial society, oriented to nature and the seasons, had little concern with the detailed passage of time, except over the scale of a year. But I find it fascinating that over 1,600 years ago, Augustine said time was something that was always coming up in conversation. This certainly continues today, as is clear from the simple reality of the word's usage in the English language. According to Oxford Dictionaries, 'time' is the 55th most common word in use today, and is *the* most common noun found in written English.

It might seem strange that I felt the need to quote a dark ages bishop on the nature of a scientific subject like time. It feels as unlikely as asking Mozart for his opinions on the benefits of using MP3 compression or lossless file types in electronic music. But, frankly, even the best modern scientists aren't much more helpful than Augustine. You'd surely expect, for example, that Stephen Hawking's *A Brief History of Time* would explain what time is and how it works. Tantalisingly, among a list of deep scientific questions that Hawking tells us have answers suggested by 'Recent breakthroughs in physics, made possible in part by fantastic new technologies', is 'What is the nature of time?' But you can search the book from end to end (and I have, so you don't have to) for any suggestion of what time is, or how it works. There is plenty on how we observe time, and how interaction with matter can change these observations, but there is nothing deeper.

Travelling through another dimension

Einstein gave us an image of time as a fourth dimension. In his special theory of relativity (coming up in the next chapter), we don't think of space and time as separate entities, but rather as spacetime – a mashup of the two. And it's easy to then imagine that time is directly equivalent to space, just a special fourth dimension that we move through at a standard rate. This is certainly the way that H.G. Wells envisaged it, pre-guessing Einstein in *The Time Machine*, published ten years before Einstein's paper, when Wells wrote:

> 'Clearly,' the Time Traveller proceeded, 'any real body
> must have extension in *four* directions: it must have

Length, Breadth, Thickness and – Duration … There are really four dimensions, three which we call the three planes of Space and, a fourth, Time. There is, however, a tendency to draw an unreal distinction between the former three dimensions and the latter …'

But spacetime is significantly more complex than simply imagining a four-dimensional block of reality that we move through at a rate of a second every second. What could 'a second every second' even mean? Some think it clearer to use a model called the block universe, where there is no movement through time – all past and present is there in the block. We merely have the illusion that time is passing. If this seems crazy, bear in mind that we don't truly experience the passage of time. It's not like watching a car pass, where you can see it approaching, passing by, then heading away into the distance. All we can be aware of is the present. We have a memory of what we think happened in the past,* we can imagine what may come in the future, but all we ever experience is the moment that is 'now'.

This struggle with the existence of the passage of time goes back at least 2,500 years. A school of early Greek philosophers, the Eleatics, considered that almost everything we associate with the passing of time, specifically change and motion, were illusions. This wasn't by any means a universal view among the Ancient Greeks. Greek philosophy from different schools

---

* You may think you have a memory of what *did* happen in the past, but all the evidence is that this is untrue. In a 1901 experiment, a class of university students witnessed a murder. The event was faked, but students did not realise this. In accounts written directly afterwards, eight different names were given for the murderer and accounts differed wildly, even down to details such as whether or not the murderer left the scene.

could be wildly contradictory. Heraclitus, a contemporary of the Eleatics, for instance, held that change was at the heart of everything, making remarks like: 'No man ever steps in the same river twice.'

The clearest example of the Eleatics' dismissal of change and its relationship to time comes through in the paradoxes dreamed up by a member of the school called Zeno. Perhaps the most apposite paradox here is one called 'the arrow'. Imagine there is an arrow suspended motionless in space and another is flying past it, shot from a bow. Let's examine the situation in the moment at which the second arrow is immediately above the first. How can we tell in that snapshot of time that one arrow is moving and the other is not? There appears to be no difference between the two in that moment, Zeno argued, so how does one arrow know to change its position in the next moment?

We now know a lot more about the nature of inertia and kinetic energy, but the paradox is still a useful one for examining what we mean by an infinitesimal moment in time – and indeed to consider whether such a thing as an instant can truly exist.

Of the two best-known Ancient Greek philosophers, Plato considered time to be a kind of unreal extension of the present into an imaginary past and future, while his pupil Aristotle linked time immutably to motion. He argued that movement was necessary for time to exist – that time was, in effect, measured by motion. Without movement, he believed, time would cease.

As far as the modern physicist is concerned, what we loosely refer to as 'time' in common experience has a number of related functions. In effect there are three different aspects of time. A first role is a particular kind of coordinate system, just as latitude and longitude give us a spatial coordinate system on the Earth.

This coordinate role seems most natural when taking the relativistic view of 'spacetime' as a whole, though spacetime isn't essential to use it.

A number of physicists have quoted the American physicist John Wheeler as the originator of the neatly pithy description of this kind of time as 'Nature's way of keeping everything from happening at once'. Just as Wheeler is often incorrectly attributed with dreaming up the term 'black hole', his description of time is also more of a quote than an original. Wheeler himself said he had seen the expression somewhere as a graffito (academics experience a better class of graffiti than the rest of us), while at the time of writing the earliest known occurrence was in a 1929 science fiction novel called *The Girl in the Golden Atom* by Raymond King Cummings, which uses the almost identical phrase: 'Time is what keeps everything from happening at once.'

Just as we need spatial coordinates to make use of space and to allocate all the entities that occupy space, we also need to allocate that picture of space to one or more time coordinates. It isn't enough to describe the world to be able to place every known entity at a particular location. We also need to know *when* this snapshot applies. Without the concept of a coordinate in time we would have no way of separating these snapshots of the universe. It's the difference between a film seen as a series of frames, each frame representing a coordinate in time, and all the frames being superimposed so there is only a single frame. It wouldn't make a very entertaining movie.

Typically, a glance at your daily schedule will reveal the use of both space and time coordinates. There's not much use saying that you are meeting friends at the Odeon Cinema in London's Leicester Square. Equally, there is little benefit in just saying you are meeting those friends at 7pm on Wednesday, 1 February

2017. In each case, the information provided pins down one type of coordinate but leaves the other unspecified. The event in the schedule needs to have both location and time coordinates before you can meet up and enjoy the film.

The phrase 'Odeon Cinema in London's Leicester Square' isn't strictly a set of spatial coordinates, but it's a label enabling us to pin down a specific location (if necessary with a little help from Google Maps) that is more humanly accessible than a true coordinate system like latitude and longitude. Similarly, if I refer to a meeting as taking place 'next Monday' it can be easier to locate mentally than would a specific date. Whichever way we choose to identify a coordinate in space or time, we are using a relativistic process. This is more obvious with 'next Monday', where the statement is made relative to the present, than it is with 'Wednesday, 1 February 2017'. But in the latter case we are also making use of an arbitrary fixed point, in this case, the year 1 in our current calendar, to establish the location of this date with respect to that fixed point.

This relative nature of time coordinates is often more explicit in the way that computer programs are written, which is why there was panic leading up to the year 2000 with predictions that the 'millennium bug' would make planes fall out of the sky and hospital equipment give up the ghost. Dates in computers are held as a number, which is often the number of days since a particular date – 1 January 1900, for instance. If the amount of space allowed in the program to store this number is limited, which it often is, there is the possibility of it resetting itself when running out of space, turning 1 January 2000 into 1 January 1900, leading to all sorts of embarrassment.

A simple example of the issues involved would be for the computer to calculate someone's age. This would be derived by subtracting their birthdate number from the current date number. But if, thanks to the bug, the current date had a smaller date number than the birthdate, the result would be a negative age. Such unexpected values could then crash the whole system. In practice, the millennium bug proved to be far less of a problem than was anticipated, but when it did turn up, it was a mistake that had its foundation in the relativistic nature of dating systems.

Unreal time

It's not uncommon for physicists to make comments along the lines that time doesn't exist, but the odd thing is that they don't really believe this. What they really mean is that many physical laws can be independent of the 'flow' of time as we perceive it and that those apparently timeless laws are all we need to predict how things will evolve. For that matter, as we will discover in the next chapter, special relativity shows that a value for elapsed time is not an absolute, but something that depends on the viewpoint of the observer. However, without time there, fulfilling its different roles, very little physics would actually be useful or in some cases exist. And while some physicists may argue that time is not fundamental, they still recognise its significance for their everyday lives in practical terms.

To see how time has such an important, if sometimes subtle, role in apparently timeless aspects of physics, think for a moment of the conservation laws. These are particular fundamentals of nature that stay unchanged in a closed system – which is just a way of saying an environment that is closed off so nothing can get in or out. Things that are conserved include, for instance, energy and electrical charge. These are important natural laws, without which science would be impossible. (Conservation laws are also, incidentally, why almost all magic is not realistic – magic nearly always appears to break one or more conservation laws.) Yet conservation is a meaningless concept without time. Conservation means that these values are the same at different time coordinates. If time were not to fulfil this function, there could be no such thing as conservation.

Following on from this coordinate role, a second role for time is to provide a measure of the 'distance' between two events that take place at different time coordinates. This provides us with one of the more common definitions people will provide when asked what time is: 'It's what clocks measure.' Let's imagine a universe full of entities, one of which changes colour at a number of time coordinates. (Traffic lights would be an example.) There is clearly a difference between a light that flashes colours like a disco light and a typical traffic light sequence. The traffic light usually stays on, say, green for a certain duration. That duration of the stay on green is the distance between two time coordinates.

Time in this sense is the measurement of a process, of the relationship between two changes. This kind of time is to a time coordinate what distance is to a spatial location. While it's easy enough to say that time is what we measure with clocks, it isn't quite so obvious what we are doing when we make such a measurement. When we measure a spatial distance we can imagine holding up a yardstick and making a direct comparison between the positions of objects and the markings on the yardstick. When we measure a distance in time we are likely to think of checking a clock at the beginning and end of the duration and calculating the difference. However, to do this reliably we need to have a clear idea what it means for the first tick of the clock to be simultaneous with the start of the event, and for the final tick of the clock to be simultaneous with the finish of the event.

It is reasonably clear what we mean when making a physical measurement and saying that one end of a ruler is collocated with the starting position, but simultaneity in time is a more fluid concept once relativity truly gets a hold. Whether or not events are simultaneous can be decidedly slippery once movement in space is also involved, which will be the next addition to our universe once time is established.

The third way we tend to look at time is as a direct equivalent to space, but one that has only a single dimension, adding a fourth dimension to reality to make up spacetime. As we have seen, this is an alluring concept that goes all the way back to H.G. Wells. And it is a very useful approach when considering the physics of the universe. However, while time can indeed be considered such a fourth dimension, it is clearly very different from the spatial dimensions.

A traditional way to illustrate the difference is to think about making a film of sand running through an hourglass. Such a movie might not win an Oscar (though it could win the Turner Prize), but it is a useful way to explore the time dimension. Let's imagine we reverse the horizontal spatial dimension by putting the film in the projector back to front, so we have swapped left and right. Leaving aside the purist's objection that it would be obvious something was changed because the emulsion would be on the wrong side of the film, so the focus would shift, it wouldn't be possible to tell just by watching whether or not the spatial dimension had been swapped around.

Now imagine reversing the time dimension by running the film through the projector backwards, from end to beginning. In this case, we would see the sand flow upwards from the 'destination' glass into the 'origin' one. The movement would look wrong; it would clearly be unnatural. We would know that time had been reversed. If we examine the block of spacetime, the different spatial dimensions have no distinction of direction, but the time dimension has a clear arrow pointing from past to future. Forwards in time is patently different from backwards.

You may have spotted a flaw in the film projector model. I chose to reverse the film in the side-to-side dimension and saw no change. But what if I had reversed it in the up-down dimension by turning the movie projector upside down? Although the sand is still flowing from the 'origin' glass to its destination, as it does in the normal situation, it would still be obvious that something weird was happening. But this is just due to our human experience of being on the Earth and thinking that the direction that gravity will pull sand is always downwards. If we had showed the whole picture, including the Earth, it would clearly be the same event whether the projector was upside down or right side up. But there is no mechanism that would make the reversal of time look natural.

It is possible to set up situations that are symmetrical in time as well as space, and in their simple models, physicists often do so. If we just see two pool balls heading towards each other, colliding, and then heading back away from each other, then the film could be run backwards and we would see no difference. But this is cheating because we are not being shown the whole picture. (This is often the problem with physicists' simple models.)

In one sense, the cheating is obvious. We know that pool balls experience friction as they cross the table, and lose energy when they collide in the form of heat loss and the sound of the collision. So the balls would be slower in the later part of the experiment, and this could be detected to indicate that the film was running backwards. Scientists are, of course, aware of this, and simply say that for the purposes of this experiment they are considering imaginary pool balls that are frictionless and lose no energy on collision.

However, there is still another problem, which has the potential to lead many scientific experiments astray. This is the risk of cherry-picking. Usually this is where experimenters choose only the results that match the outcome they want, excluding results or whole experiments where the outcome runs counter to expectations. Sometimes this cherry-picking effect can happen without the experimenters even being aware that they are doing it.

Take, for instance, experiments that have been undertaken in the past to see if telepathy existed. A group of people were tested first, and those with the best results were chosen for further experimentation. This kind of experiment is very time-consuming, so to maximise the amount of data available, the scientists included the data from these successful selection tests in the main body of data, as these were exactly the same test as was later used. It seemed a waste of valuable data just to throw it away. And the outcome was to detect significant, if not very strong, evidence that telepathy did exist.

However, by using the data from the selection tests, the scientists were unconsciously cherry-picking. Imagine that there was no telepathy and that the results were just a matter of random guessing. If we add together the results from all the selection tests and all the later tests, then the outcome should show no evidence for telepathy. But if we only make use of the selection tests where the candidates scored highly – which would happen because only those candidates were allowed to go forward – it will bias the overall results. And this is exactly what happened.

Although it's less obvious, cherry-picking is also taking place in the pool ball experiment. Balls do not suddenly, of their own volition, begin hurtling towards each other across a pool table. The actual experiment includes giving the balls a push to get them moving in the first place. But those frames of the movie have been removed, cherry-picking only those frames where the experiment does indeed appear to be symmetrical in time.

If we were to include the entire process, so we also see the point where the balls are given a push to get them moving in the first place, there is still a clear direction in time that would become distorted were the film to be played backwards. Time has that elusive arrow, a clear natural direction between the time coordinates in which events occur in the universe.

However, while it is hard to argue that the arrow is not there, there is certainly a problem with the concept of the 'now' moving through spacetime at a steady pace (relativity permitting) on the time axis of the four-dimensional block. Or, for that matter, with time passing by, like the poet's ever-rolling stream that bears all its sons away. This is because motion is measured as the distance that has been travelled along a spatial axis in a second. But if we consider travelling along the time axis, in the direction of that arrow of time, we end up moving at the rate of a second per second – and that self-referential aspect is, to say the least, uncomfortable.

The universe from outside

Let's go back, then, to the block universe, imagining the whole of spacetime as a four-dimensional block. For convenience of imagining (because four dimensions are something we struggle to get our heads around), we can think of spacetime as three-dimensional with two spatial dimensions and one in time, ignoring one of the space dimensions. Being able to look at that block would be impossible for us in the real universe because we are, by definition, of the universe – inside it. However, the great thing about thought experiments is that physical constraints cease to be a problem.

So let's take the 'God view' of the spacetime block, looking at the whole thing from the outside. The two spatial dimensions take in the entirety of space in the universe, whether that happens to be finite or infinite (in the latter, the God view is rather strange, but not totally inconceivable), while the time dimension stretches back to the big bang, or even further if one of the other versions of cosmology is correct, and off towards what could be an infinite future. As we scan along the time dimension, we see the universe develop through all its different forms.

Zooming in on the space dimensions that encompass our solar system, by following the time dimension we see the solar system form gradually until at a point we would regard to be around 4.5 billion years in the past, the Sun ignites, after which we can watch the Earth go through its many changes. One thing we can't do, though, is find a label in the block universe saying: 'You are here, this is now.' There is no present in the block universe. There is no point along the time dimension that is specially privileged: no past and no future. There is no now. Just all of time, laid out before us.

However, time's arrow still exists even in this view. One direction through the block along the time axis is very different from the opposite direction. Many simple physical processes suggest that this shouldn't be so. They are totally symmetric in terms of time with no requirement for a particular direction to be taken. (Though to make them simple, we usually have to cheat, like the physicists do with the colliding pool balls. The real world is rarely simple.) But for the main source of time's arrow in much of physics we have to look to a surprisingly straightforward and workaday aspect of physics called thermodynamics, originally inspired by the need to improve the effectiveness of steam engines.

As the name suggests, thermodynamics was developed to describe the way that heat flows from place to place, and the key aspect of it that is so important for time's arrow is the second law of thermodynamics. This can be described in two ways: that in a closed system, heat always flows from a hotter to a colder body; or that the entropy in a closed system will remain the same or increase.

This 'entropy' is a measure of the disorder in the system. The law is saying that the level of disorder will stay the same or become greater. Entropy is not just a vague concept like 'disorder' but has specific mathematical values depending on the way the components of a system can be arranged. The more different ways there are to achieve the same outcome, the higher the entropy and this disorder. So, for instance, there is only one way to arrange a set of books in alphabetical order (if there are rules to cover oddities like two books with the same title) – so this has very low entropy. There are considerably more ways to arrange the books so that all the books beginning with the same letter are adjacent, so this has more entropy. And there are lots of ways to arrange the books totally randomly – so this has high entropy.

This may seem very different from 'heat flows from a hotter to a colder body'. The book example is clearly not about temperature. But imagine we have a separate hot and cold body. This system has low entropy, because all the hot, faster-moving atoms are in the hot body and all the cold, slow-moving atoms in the cold body. When we bring them into contact, some of the fast-moving hot atoms bump against the slow-moving cold atoms. The hot atoms will slow down and the cold atoms speed up. Now some of the atoms in the hotter body are cool, and some of the atoms in the colder body are warm. As with randomly organising the books, there are significantly more ways to arrange this, so the entropy has increased. There is more disorder, with a mix of speeds of atoms in both bodies.

Reversing entropy

A knee-jerk reaction to hearing the second law can be to dismiss it as being clearly unrealistic. After all, a fridge takes heat from the inside of the fridge and pushes it out into the warmer air around it, apparently running counter to the second law. And we have pretty clear examples all around us in the natural world of the extremely ordered structures of living things emerging from the disorder of the raw materials that are consumed to make them. Even this book, in its small way, is an example of order created from disorder. Imagine the virtual pages of the book before it existed, consisting of a whole jumble of letters that have now been rearranged in the one, specific and hopefully ordered way that spells out the words you are reading and that, give or take a typo, is how my brain intended them to be. That's a massive reduction in entropy.

However, all these situations manage to arise despite the second law of thermodynamics because of a get-out clause that I slipped quietly into my original description of that law. I said that the law applies in 'a closed system'. This is in reality a huge cop-out, a way of excluding from the law pretty well everything that really happens in the universe, as we hardly ever experience closed systems. A closed system is one in which nothing, particularly not energy, travels in or out.* And, with the possible exception of the universe as a whole (and even there we're not entirely sure), this just doesn't happen.

In the specific examples I raised, the fridge takes in electrical energy from the wall socket to power its ability to overcome the second law and pump heat from the cold inside to the warmer exterior. As for the Earth, it has a vast amount of energy pouring into its system all the time from the Sun – around 89 billion megawatts of our friendly neighbourhood star's output hits our planet. Without this energy there would be no life on Earth. And even the production of this book, the ordering of the letters to produce the words you are reading, took the energy my brain consumes, totalling around 20 per cent of the energy consumption of the resting human body, plus the energy involved in the physical effort of typing, editing, printing and distributing it.

There is one other significant oddity with the second law of thermodynamics, something that wasn't fully appreciated for some decades, which is that the law is statistical rather than absolute. If we imagine a simple model consisting of two closed boxes full of gas, one hot and one cold, which we join together to form a single box isolated from the world, we would expect, according to the law, heat to flow from the hot box to the cooler

---

* By the definition I'm using of a closed system. An alternative definition calls this an isolated system, and a closed system prevents matter but not energy from travelling in and out.

one, ending up with the joined boxes equalising at a mid-way temperature.

If we look at this from the point of view of heat flow, it makes good sense. Temperature is a measure of the average energy of the atoms or molecules that make up a substance. In the hot box, the atoms would be whizzing around faster than those in the cold box – that's what being hot means. Once the gases start to mix, there is no longer one box of fast atoms and one of slow – instead we will end up with both fast and slow atoms in both boxes. Each side will take on an intermediate temperature, eventually roughly equalising.

However, this outcome does depend on what all those individual atoms do. In principle, because we are talking about a collection of random occurrences, we could discover an experiment where we then went from having the same temperature in both boxes to one box becoming hotter than the other. This could happen because purely by chance more fast atoms (say) went in the left-hand box and more slow ones in the right-hand box.

It can be clearer to imagine this happening by starting with boxes that just have two gas atoms each. To begin with, both hot atoms might be in the right box and both cold atoms in the left. After a while, it's likely that you might have one of both types of atom in each box. But after a little more zipping about, you could end up with both hot atoms in the left-hand box and both cold in the right. That's just how such random occurrences stack up.

With the vast number of atoms that would be in real boxes of gas at atmospheric pressure, such a separation is extremely unlikely to happen. It's as if the same lottery numbers came up week after week. But run the lottery often enough and you will get this kind of sequence. It is admittedly very, very (very) unlikely, but with enough repetition, the unlikely configurations will turn up. And the same is true of the second law. It is not an absolute law, but one that applies relative to the size of the population and the likelihood of a particular configuration occurring.

Scrambling letters

Entropy can be calculated by looking at the different ways it is possible to arrange the items in the system. The more ways that the components can be arranged, the higher the entropy (aka disorder) is. Thinking about the contents of this book, it contains a total of somewhat over 500,000 characters if we include spaces. The number of different ways to arrange those half a million characters is astronomical. Mathematically it's described as 500,000! or 500,000 factorial, meaning 500,000 × 499,999 × 499,998 × 499,997 ... If you think about it, there are 500,000 places in which we could put the first character, 499,999 places to put the second character once the first is already in place, and so on.

Not surprisingly, 500,000! is an immense number. My calculator gives up the ghost and says 'overflow'. An online factorial calculator I tried simply came up with 'infinity'. This isn't true – the result is finite, but mind-bogglingly large: far bigger than the number of atoms in the universe. (Which is why a roomful of monkeys are not in practice going to type out the works of Shakespeare any time soon.) A better online calculator I found used an approximation method to come up with $1.022801584 \times 10^{2632341}$ – just think about that for a moment. That $10^{2632341}$ is 1 with 2,632,341 zeroes after it. Just to write out that number would take over five books of this length.

So that's how many ways we can arrange the letters and spaces that are present in this book. How many ways are there to make this exact and specific book? One. Admittedly that's only in the strictest sense of 'this exact and specific book', where each letter 'a', for instance, is different from each other letter 'a' – imagine each of the 500,000 characters has a serial number. In reality, they're not functionally different. Let me share a secret with you. After I had written the entire book, I swapped the two letter 't's in the word 'written' earlier in this sentence. So the book you are reading does not have the same arrangement of letters I first used – but no one would have noticed this if I hadn't mentioned it.

There are lots of ways to perform this kind of permutation – swapping letters and still getting the same set of words – but the outcome is still a tiny number compared with $1.022801584 \times 10^{2632341}$. So there are far fewer ways to arrange the letters to make this book than there are to arrange them without them forming my words. And that means that the content of this book is far less disordered than an unstructured collection of those letters. Which is just as well if it is to contain any information.

If the letters in the book were loose, so they could fall out if you dropped the book (not a happy thought), then we would see time's arrow in action. If we watched a film of this occurrence it would be easy to say which way the film was being run. If it went from being the actual book to an incomprehensible mess of letters, the film would be running forwards. If it went from a mess to a readable book, the movie would be running backwards. Time's arrow either keeps things the same or leads us to greater disorder, greater entropy.

The end of time

Arguably there is a position we could imagine where time manages to lose its entropy arrow. If we consider the universe to be a closed system and the inevitable future of the universe to be the increase of disorder overall, it must eventually approach a state of maximum disorder. Total chaos. If that were to happen, then there would no longer be a clear distinction between forwards and backwards in time because neither direction would result in an increase in entropy. All entropy could possibly do is decrease. However, such a near-equilibrium state is such a vast distance into the future that it has no effect on our current understanding of time and time's arrow.

This necessarily puts time on a different footing to the other components of spacetime. There is no reason for picking one spatial direction over another – they all have equal weight. There is no 'arrow of space'. But the time dimension comes with the added bonus of a clear pointer saying 'this is the way that things progress'. As we have seen, this isn't always obvious. There are some physical processes, like the central section of the colliding pool balls movie, where it appears that you could run the movie backwards without any distinction.

Physicist Sean Carroll has argued that the lack of an arrow of time in these simple processes indicates that symmetry in time is the 'natural' state of affairs, but because we are relatively near the beginning of the universe (a mere 13.8 billion years ago), he suggests that we are under the influence of this odd state of affairs where entropy is still very low, rather in the same way that being near the Earth appears to give us a special direction of up and down, where there isn't really one.

However, such apparently symmetrical physical processes, though easy to pick out in simplified models where, as we have seen, cherry-picking is applied, are pretty well impossible to identify in the real universe – perhaps demonstrating that they are more a phenomenon of the way these limited models are constructed, rather than a result of a hidden symmetry in time in a universe where entropy reigns.

Carroll argues that the influence of the arrow of time is misleading. He says: 'Our unequal treatment of past and future is a form of temporal chauvinism, which can be hard to eradicate from our mind-set. But that chauvinism, like so many others, has no justification in the laws of the universe … it is a mistake to prejudice our explanations by placing the past and future on unequal footings. The explanations we seek should ultimately be timeless.' But to say 'explanations should be timeless' is surely itself chauvinistic – Carroll is looking for a solution that fits with his particular worldview. In a universe where the arrow of time makes itself felt so plainly, it seems perverse to put so much effort into pretending that it's not there.

However, leaving aside the obsession some physicists have with doing away with time, the reason Carroll is discussing this possibility is to talk about cosmology as an explanation for the source of the arrow of time, which makes a lot more sense. As he points out, we believe that originally, around the time of the big bang, the universe was very simple, with an almost total lack of structure and almost total order.

As is often the case with entropy, this isn't an obvious assertion. In the examples we've used so far, like the hot and cold objects and the letters in this book, a lack of structure resulted in disorder, not order. When the hot and cold objects are in contact, reducing structure and order, there is an increase of entropy. However, the very early universe is assumed to have been in an unusual state where there was no distinction between its components, so there isn't a sense of disorder from the mix. It's a bit like a book in which every letter is the same. It might not contain any information, but equally it can't be considered to be disordered even if those letters are randomly arranged.

We don't know why the early universe should have been in such a low state of entropy, though some models suggest that it was because some or all of a pre-big bang universe went through a cataclysmic process that wiped out previous disorder and complexity, leaving it smooth and featureless. However, the extremely low entropy of the early universe combined with the second law of thermodynamics make that arrow of time an inevitable outcome. And this has resulted, indirectly, in the eventual formation of solar systems and planets, life and humanity.

Not all scientists agree with Carroll on the need to discover timeless solutions or even that there is anything sensible in the search for them. Another eminent physicist, Lee Smolin, suggests that the reason physicists tend to push time aside is that even they, despite being immersed in relativity, feel an urge to fight back to the absolute. It has been traditional, Smolin argues, to think of our most valued concepts – truth, love, God – as being outside of time. True absolutes. Scientists know better when it comes to gradually removing the absolute from our small-scale views of the universe, but arguably still feel this craving to escape relativity. And it comes through in the assertion that physics should operate outside the influence of time.

This desire is, in part, a convenience. Physicists don't want the laws of physics, for example, to change with time, because that would make life difficult for them. However, Smolin argues, it is perfectly reasonable to think of time as real. He envisages a set of real moments that are 'now'. In his picture, the past is not real, but has an influence on the present, so that we can examine and analyse data from the past, while the future has no existence and no 'echo', so is open and never completely predictable. Laws of nature, says Smolin, are not timeless, but are features of the present, and they can evolve.

Part of the problem we have here, Smolin suggests, is that mathematical structures like numbers and curves are timeless, and the more familiar we get with the ability of mathematics to predict and approximate to the physical universe, the easier it is to be fooled into thinking that the universe should operate without a dependence on time too. This is not just a philosophical issue. Smolin comments: 'The dream of transcendence has a fatal flaw at its core, related to its claim to explain the time-bound by the timeless. Because we have no physical access to the imagined timeless world, sooner or later we'll find ourselves just making things up.' We will return to Smolin's viewpoint as we bring in motion, where the existence or otherwise of time becomes crucial.

Subjective time

What certainly is true is that the way we as humans experience time is purely relativistic. Simple experience tells us that subjective time has a remarkable elasticity. Minutes can stretch to hours. Days can flash by like seconds. Einstein infamously claims to have undertaken an experiment on the subjective nature of time. I have frequently seen his 'paper' on the subject quoted as if it were a real publication, though the acronym formed by the alleged publication it appeared in, the *Journal of Exothermic Science and Technology*, makes this seem unlikely.

The subject of Einstein's experiment was supposedly himself, in a process undertaken with the help of silent film star Paulette Goddard, whom Einstein had met through their mutual friend, Charlie Chaplin. Einstein summed up the 'experiment' in his abstract: 'When a man sits with a pretty girl for an hour, it seems like a minute. But let him sit on a hot stove for a minute and it's longer than any hour. That's relativity.'

Einstein nevertheless had a serious point to make about the difference between the measurement of time that we experience subjectively and the measurement of time used as the basis for data produced by scientific experiments. Although we can train ourselves to count seconds reasonably well, subjective experience of the passage of time is hugely influenced by what we are doing during the process. And in that sense, the rate at which we experience time passing is a strongly relative phenomenon.

But there is more to time's relative nature than our internal experience. This 'relativity' is not what is usually meant by relativity in the scientific sense. As far as Isaac Newton was concerned, from the scientific viewpoint, relative time was nothing more than a means of generating units of measurement. In his masterwork *The Principia*, Newton writes:

> Absolute, true, and mathematical time, in and of itself and of its own nature, without reference to anything external, flows uniformly and by another name is called duration. Relative, apparent, and common time is any sensible and external measure (precise or imprecise) of a duration by means of motion; such a measure – for example an hour, a day, a month, a year – is commonly used instead of true time.

However, this view was disputed from the beginning. Newton's contemporary Gottfried Wilhelm Leibniz argued that since God created relative time with the universe, and there was no rational reason for beginning everything at any particular point in any absolute timeframe that was external to the universe, that this relative time was all there was.

For Newton, on the other hand, time and space remained absolute. Relativity came from the way that we made measurements by, for instance, tracking the position of something that was in motion, whether it was the Sun in the sky or the hand of a clock. Newton's view would be rendered an illusion by Einstein. His special theory of relativity tells us that time as observed by objective physical experiments has nothing of the absolute about it. Depending on where we observe it from, time can be made to run fast or slow. But before we encounter this theory, there is one more component required.

With matter, space and time assembled, we can take the next step in constructing a relativistic universe. Once we have these components, motion is possible.

5 Motion

◇◇

We live in a universe of movement. At the level of fundamental particles nothing is ever still. The only apparent absolute in terms of motion is at the temperature absolute zero, where all motion should stop, but this is unreachable. Though Zeno thought that he had disproved motion's existence, his paradoxes *are* paradoxes because motion is so obviously a part of everyday life. And with one exception, motion is in the hands of relativity.

This wasn't obvious to begin with because the Earth is too big and too near to the things that we experience every day. Because of Earth's looming, unavoidable presence, our planet provides an apparent absolute, immobile frame of reference, making it easy even today to misunderstand just how relative motion really is. When someone says they are moving or still, they will almost always really be describing their motion with respect to the Earth.

Yet we also have experience in everyday life that demonstrates relativity in action. We expect two cars heading towards each other to approach one another faster than either individual's speed and to crash at their combined speed. We expect, should we run alongside another runner whose velocity is the same as our own, to keep alongside – not to move with respect to each other. Because it's not the Earth's reference frame that matters here, it is our own. This is the Galilean relativity we met on page 1 in Galileo's boat on Lake Piediluco, the first of the great relativistic observations of science, providing the birth of modern physics.

Think about your current situation as you read this book. Are you stationary or are you moving? You might be on your couch at home, meaning that you are still, or on a train or in a plane, and hence moving. But that description is the result of the Earth fooling us once again by imposing a single frame of reference. When you think that you are sitting still, you are rotating with the Earth's spin, orbiting the Sun and moving with the solar system around the Milky Way galaxy. Your velocity is arbitrary. It is all a matter of what is used as a reference frame for your motion.

Does the Earth go around the Sun or the Sun around the Earth? It appears clear from naive observation that it is the Sun that moves around us daily. Usually, it is more convenient to consider that the Sun's daily movement through the skies is caused by the Earth's rotation. Yet bear in mind that this 'correct' interpretation is merely the one that makes the calculations work most easily. It is the one based on a frame of reference detached from the Earth's surface. In terms of pure motion, it is entirely reasonable to consider the surface of the Earth fixed and the universe, including the Sun, in daily rotation. It really is all relative, depending on the frame of reference you choose to observe. (Having said this, there are some rotational effects that need a more complex explanation – but we'll come back to that.)

Newton's bucket

Once we have motion in our model universe, Newton's laws give us an independent view, a way of abstracting ourselves from the Earth-centred misunderstanding of the Greeks. As we have seen, Newton explicitly made use of absolute time and absolute space in his *Principia*. Although he was well aware of the existence of relative space, where our motion is measured with respect to some other entity in space, he also believed that there had to be an absolute and fixed space. The origin of his belief was probably religious – that God, if you like, provided that absolute source of reference – but more frequently the absolute frame would be provided by the ether, and Newton did have a scientific argument for an absolute, based on an oddity observed when things rotate.

Newton's favourite illustration of this 'proof' of the existence of absolute space involved a bucket of water, of which more in a moment, but a simpler and more personal approach is to think what happens when we ourselves rotate. When we spin around, we feel dizzy (and this is as true of an astronaut in space as it is for us on Earth, so we can't blame this on the gravitational pull of our planet, though Newton did not know this). But this effect occurs whether or not other things are rotating with us. So what is our rotation causing that dizziness measured relative to, if not to some absolute grid of space that remains fixed?

It's fair to say that this whole area remains to be fully settled in the minds of physicists. Rotation is very different from movement in a straight line, because it features acceleration. (Acceleration is a change in velocity, a value that has both a speed and direction component. Steady rotation involves constant change of direction, and hence constant acceleration.)

While it is impossible, as Galilean relativity makes clear, to tell inside an enclosed ship whether that ship is standing still or moving at a steady speed, it's very easy to tell if you are accelerating. In the case of rotation, apart from anything else, dizziness tells you.

There is also, however, an indication with a physical object outside the body, like Newton's bucket of water, which acts differently if the object is spinning. Whirl a bucket around, and the water rises at the edges of the bucket. Fix the bucket to the floor of Galileo's enclosed ship and spin the ship and the water will still rise. You can see this rise of the water inside the ship, even though in your ship-bound frame of reference the bucket is not spinning, it's stationary.

Some argue that this is still a relativistic effect, rather than a reflection of an absolute fixed reference point that is not spinning. According to a nineteenth-century idea called Mach's principle, the rotation should be measured with respect to the whole of the rest of the universe, and you would get exactly the same effect as rotating a spaceship, say, including dizziness, if you kept the spaceship still and rotated the whole universe around it at the same rate. But, as Richard Feynman once observed: 'Well, I do not know what would happen if you were to turn the whole universe, and we have at the moment no way to tell. Nor, at the moment, do we have any theory that describes the influence of a galaxy on things here so that it comes out of this theory … that the effect of rotation, the fact that a spinning bucket of water has a concave surface, is the result of a force from the objects around.'

This aspect of spinning has not been securely solved, but it is one that can be largely ignored for the purposes of exploring motion. We know what happens, and are able to make use of it, even though we can't explain why it happens, just as we aren't able to explain why the electron and the proton have the same magnitude of charge. However, this doesn't mean that we can avoid rotation. Because if there's one thing that bodies in the universe like to do, it's spin. And it's just as well. Without things spinning around we wouldn't exist. It's the spin of the material that condensed to form the solar system that enabled it to form a stable structure with planets, essential for life to evolve.

Angular momentum rules

It might not seem obvious why pretty much everything spins, but it comes down to contraction. Whether it's a star and its solar system forming, or a whole galaxy, the process that makes it all happen is a vast, diffuse collection of gas and dust gradually pulling together under the influence of gravity. Let's assume that the original material is rotating ever so slightly (we'll come back to why this is likely to be the case in a moment). Then, as it pulls together, there will be an irresistible implication that the rotation will speed up.

This is because of the conservation of angular momentum, one of nature's key conservation laws. The law effectively says that the 'oomph' with which something rotates stays the same. The further out a mass is from the centre of rotation, the more angular momentum it has for any particular speed of rotation. So if a rotating mass is pulled in towards the centre it has to rotate faster in order to keep the angular momentum the same.

This is something that is inevitably compared to a spinning ice skater. If an ice skater goes into a spin with her arms sticking out horizontally, then pulls her arms in close to her body, her spin speeds up to a remarkable degree. As it happens, my daughters had ice skating lessons when they were younger and I've seen this many times. But if you don't frequent a skating rink, you can see the same effect if you go on one of those pieces of playground apparatus where you hang onto a stand that rotates on a pole. Get some rotation going leaning outwards and you will speed up impressively when you pull yourself in towards the pole.

So a small amount of rotation when a huge cloud of gas and dust is spread over a wide area will become a significantly faster rotation when it pulls in to form a star or the elements of a solar system. (Such systems tend to form a disc rather the way a spinning piece of dough ends up as a pizza, because there are forces in the plane of rotation that aren't present at 90 degrees to it. It's only when you get a concentrated enough collection of matter like a planet or a star that gravity can dominate and produce a roughly spherical shape, and even then there will be a bulge around the middle.)

So far so good, but where did the small amount of rotation come from in the first place? In a perfectly evenly distributed, static spherical cloud of gas and dust, it would be possible for the whole thing to contract without there being any rotation. But the real universe isn't like that. There's no particular reason why the initial cloud that collapsed should be a perfectly uniform sphere. In reality it is going to be a messy shape with varying density, influenced by all kinds of other objects around it. So as it contracts, there will be more stuff on one side than another, making for slightly more gravitational attraction in one direction than another. (It's also possible there will be forces in a particular direction due to electromagnetism, if the cloud is ionised.) The asymmetry of the material will mean that some bits are pulled more than others – resulting in the particles not all heading straight for the centre of mass, but moving sideways as they move inwards. Net result: a small rotation that can then be amplified by the contraction.

Motion emerges

Assuming that, unlike Newton, we are comfortable with relative space and time, we can start to analyse motion – the way that the position of an entity changes at different points in time. Here's where it's easy to see susceptible scientists slipping into the trap that Lee Smolin identifies of confusing timeless mathematics with time-embedded reality.

A common enough activity for a scientist is to take a sequence of readings, providing data that will help describe the behaviour of an object or experiment. Those data points might be simply position and time of a moving object as compared to a fixed point and a clock on the surface of the Earth. We end up with a series of data points, or a graph where we plot out the motion. That end point is purely mathematical and static. It does not change. So, if we make the mistake of thinking that the model – the set of numbers or the graph – *is* reality, it is all too easy to consider that time is an unnecessary adjunct to the physical world. However, reality is quite different. The moving object is not a neat set of numbers or a curve, nor do the numbers apply to its entire flight in existence, merely providing the data that applies on an instant in any particular moment.

However, in the Garden of Eden of relativistic motion there is a serpent – and that is light. Light forms its own absolute. It won't conform to the relativity of matter – though we were happy to characterise it as stuff, light certainly isn't matter. And that simple difference is enough to produce the remarkable implications of Einstein's first great contribution to relativity, the special theory. This threw away the absolute backdrop of Newton that could be represented by the ether (or God) and left only true relative motion. Here is the reason why spacetime becomes flexible, malleable. Bring light into the picture and Newton's insights lose accuracy with the inevitable relativistic transformation of space, time and mass.

Relativity becomes special

The special theory of relativity formally emerged in a paper written by Einstein in 1905, when he was working as a clerk in the Swiss patent office in Bern. The basic elements that he brought together had been swirling around for some time, and were present in part in the work of several others, but it was Einstein who crystallised them in a single paper. Like Newton and his apple, there is a story (or stories) to accompany the mental leap that brought Einstein to special relativity. Here there is no doubt that most of the tales are apocryphal, as there are several conflicting versions.

Since we are dealing with stories here, it's fair to select the one I like best, which is of a young Einstein lying on a grassy bank in a park, letting sunlight filter through his eyelashes. He imagined that the glittering flashes of light caused by his eyelashes were sunbeams, and that somehow he was able to ride along beside one of these luminous beams. What would it look like? Einstein derived his answer from the work of the Scottish physicist who would remain one of his heroes for life: James Clerk Maxwell.

As we saw on page 50, a few decades earlier Maxwell had deduced the nature of light by observing that an electrical wave, travelling at just the right speed, should produce a magnetic wave, which also travelling at that speed would produce an electrical wave and so on. Provided the waves went at the right speed they would be self-sustaining. And that speed happened to be the speed of light. However, this theoretical piece of intuition, borne out by later experiments, had another implication. Light had to travel at this particular speed to be able to exist.

As Einstein imagined floating alongside the sunbeam he knew that, just like Galileo and the key, in this frame of reference the light wasn't moving. However, this was a serious problem. Because without that motion, the light couldn't exist. It wouldn't be there. If light were like ordinary matter, and speeds added or subtracted in standard Galilean fashion whenever anything moved, all the light around the moving Einstein would disappear. Only in a totally static universe where all matter was frozen in space would light continue to exist.

The real world was clearly not like this. Thankfully, light does not disappear every time anything moves. Something had to be wrong with the picture. And Einstein decided to see what would happen if that 'something' was the assumption that light was like everything else and its speed changed relative to a moving body. What if, instead, light ignored the frame of reference of an observer and always travelled at the same speed, however anything moved with respect to it?* This would clearly allow light to exist – a good start. But like any Faustian bargain, something would have to be given up to allow this to occur.

---

* There was one proviso – that the observer's frame of reference wasn't accelerating. This would make things a lot more complicated, as we will discover in the next chapter.

Relativistic keys

We can see exactly how the most dramatic effect of special relativity occurs, making the flow of time itself relative, by going back to Galileo's experiment with the key in the boat and bringing it into the modern day. We need a clock that can be seen from both a moving platform and a fixed one. (Relativity makes it clear that nothing is truly fixed, but we are defining 'moving' and 'fixed' from the point of view of a particular observer.) So we're going to make a clock based on Galileo's key, which could be seen in the boat and also from the shore, assuming a good pair of binoculars.

We start by getting Galileo to throw his key up in the air with regular timing (for ease, we'll introduce a machine to do this). The movement of the key can provide us with the ticking of a clock, a measure of the passage of time. We could deal with Galileo's actual situation where the key decelerates as it travels upwards, then accelerates as it travels downwards, but even though the mathematics is within the scope of a high school student, that makes life a trifle more complicated, so it would be better to do something Galileo wouldn't have thought of – we can simplify the situation by taking the boat into space.

So now what we've got is Galileo on the spaceship *Piediluco*, travelling through space at a constant speed. To replace Galileo doing the throwing and gravity returning the key, we've constructed a special device that has two parts, one on the floor and one on the ceiling. The key starts at the bottom device and is pushed upwards so it heads towards the ceiling at a steady pace. When it reaches the ceiling, the device there pushes the key back down towards the floor at the same speed. With no gravity to provide acceleration, the key continues on its path at this constant speed.

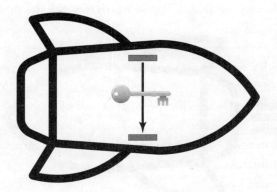

Figure 1: The key clock inside the ship.

So now we have an unusual kind of clock. Instead of a pendulum or a watch mechanism, this clock consists of the pushers and the key. And each time the key reaches floor or ceiling it provides us with a tick of the clock. Just as Galileo predicted, with the spaceship moving at a constant speed, there is no way to tell inside the ship that it is in motion. In the ship's frame of reference, the key will continue to travel up and down in a straight line whatever the speed of the ship is, as long as the ship doesn't accelerate. The passage of time, as measured by our key clock inside the ship, is not influenced by the ship's motion. This is no surprise.

Let's now do our equivalent of watching the key from the bank of the lake. This time we're watching through super binoculars from the Earth – the place with respect to which the ship is moving at constant speed. What do we see as we watch the key through the transparent walls of the ship? Let's say the key is just setting off from the top pusher. By the time the key reaches the bottom pusher, the ship will have moved. So instead of travelling straight up and down, from our Earth observer's viewpoint, the key will have to follow a diagonal path to reach the pusher.

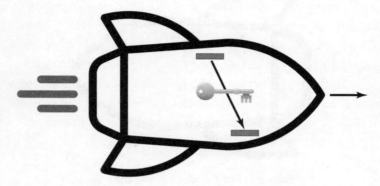

Figure 2: The key clock seen from Earth.

Clearly this involves making a longer journey than going straight up and down. So does this mean that in our external frame of reference, from the Earth observer's viewpoint, the key takes longer to make the trip? No. The reason is that we know that if two things are moving, we add their speeds together. As the ship is moving at right angles to the key, the values that come out of that addition are a little more complicated than straight A + B, involving a spot of Pythagorean geometry, but the speed of the key *as seen from Earth* will definitely be increased by the motion of the ship. And, by a happy coincidence, it is increased by just the right amount to enable it to cover the extra distance in exactly the same time as it takes to do the straight-line journey as seen from within the ship. Using Galilean relativity, time will tick on at the same rate on the ship, whether seen from inside the *Piediluco* or from Earth.

The light clock

However, Galileo didn't know what Einstein would eventually work out about light. So to ensure that his picture holds up for Einstein as well as Galileo, we need to try the same experiment, but instead of the key, we substitute a beam of light. This is a simpler clock to build, as instead of special pushers on floor and ceiling we can use mirrors. Once more, from the point of view of Einstein on the ship, the light makes a constant tick as it travels vertically up and down between the mirrors and provides our clock. But what does the observer on the Earth see?

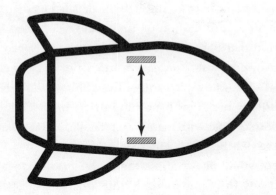

Figure 3: The light clock seen from the ship.

As before, we have to cope with the situation that during the passage of the light between floor and ceiling, the ship has moved on, so the light has to follow a diagonal path to reach the next mirror. The light is taking a longer path as seen from the Earth. But now we have a fundamental difference from the key clock. According to Einstein, the light always goes at the same speed, whichever frame of reference we choose. So we don't add on the speed of the ship. The light is still travelling at the same speed when seen from Earth as it is from within the ship.

Einstein found a way to cope with this. And it resulted in making an uncomfortable compromise. If time were passing more slowly on the ship *as seen from Earth* than it was as seen by Einstein in the ship, the light would have a chance to catch up. So on the ship everything behaves in a perfectly normal way. But when we look at the moving ship from the Earth, time is passing more slowly on the ship. Special relativity tells us that every time something moves, time slows down when seen from a fixed observer's viewpoint.

There is no need to go through the detailed mathematics that show the impact of motion on time and space, given Einstein's assumption that light always travels at the same speed, but the maths involved is just basic algebra and a spot of geometry, quite a surprise given the reputation of Einstein's work for being difficult and obscure. This is nothing that a high school student who is comfortable with maths can't handle, and it clearly shows how these strange effects emerge.

If you would like to dip a toe in the mathematical waters, the detailed thinking behind time dilation in special relativity is explored in the Appendix at the end of the book (page 279). Using the light clock as an example, it shows how the time-distorting effects of relativity follow from light's constant speed in less than a page of mathematical working.

The calculation of the size of the effect involves the ratio of the squared speeds of the object and that of light, which means that for most moving objects, travelling far slower than light, this effect is pretty well impossible to detect. But it is there and has now been measured many times. And if the ship were to travel at a sizeable percentage of the speed of light, then this slowing of time would become very obvious. It's important to realise that this isn't just an optical illusion produced by viewing the ship from a distant fixed point. It doesn't just *look as if* time is running slower on the ship. It really is, from the Earth's frame of reference. But inside the ship, which isn't moving as far as the astronauts are concerned, time is running perfectly normally. When the travellers look at the Earth, which to them is the thing doing the moving, they will see time running slowly on Earth.

The real time machine

A famous thought experiment brings home the reality of this effect. Imagine we have a pair of twins, both 25, one of whom goes off on a high-speed journey through space, and the other stays back on Earth. If the ship goes fast enough, after, say, five years' travelling, the astronaut twin who arrives home aged 30 might find that her stay-at-home twin is now 40, or 50, or long dead – depending on just how fast the ship had travelled.

An observant reader should spot a problem with this assertion. It's true that the ship's clocks run slow from the Earth viewpoint, but from the point of view of the astronaut, time is running slowly on the Earth. The effect is symmetrical. The get-out clause is that basic special relativity applies only in situations where there is steady movement – not when there's acceleration. And in the case of the astronaut, the ship is accelerated up to high speed, travels away, then decelerates and accelerates up to speed back towards the Earth. The ship has undergone acceleration as a result of the force applied to it by its engines. But there was no equivalent force applied to the

Earth. And it's this breaking of the symmetry of the situation that means, in effect, the clocks are reset so that five elapsed years experienced by the astronaut can be much longer for those left behind on the Earth.

This means that the astronaut has become a time traveller. If, for instance, the stay-at-home twin is 50 when the astronaut gets home, then the astronaut has travelled twenty years into the future. This is genuine time travel – the real, relative thing. Fictional time travel is almost always portrayed as absolute. The time machine disappears from one time and appears in another. But true time travel is always relativistic. The time machine is put into a state where time flows differently for it than it does at its destination. So when the machine arrives, the time at the destination is different from the time in the time machine. But the time ship never disappears or appears, TARDIS-style. It is always visible, but visible with a slow-running clock.

This mind-boggling time travel effect of relativity is predicted by Einstein's theory, but has also been widely shown to be the case by experiment. This was first done by flying atomic clocks around the Earth and seeing how the movement influenced the passage of time on board. The effect is tiny – but it's there. If, for instance, you flew across the Atlantic once a week for 40 years, you would move around a 1,000th of a second into the future. Similarly, the effect has to be allowed for to make satellite navigation systems work.

The satellites that form the basis of GPS are effectively just very accurate radio transmitter clocks, pinging out the time. The receiver in your car picks up the time from a number of clocks, and by the slight differences in the time signal received, works out how far away the receiver is from the satellites, giving its position on Earth. But those satellites are moving, so their clocks run slow. The shift has to be compensated for. (There is another effect from gravity that makes the clocks run fast, so it's actually the combined effect that has to be dealt with.) Our best time machine to date is the Voyager 1 probe, sent out towards the far reaches of the solar system in the 1970s, which has been travelling at a good speed for long enough to have moved 1.1 seconds into the future.

It's not just time

I have concentrated on the impact of special relativity on time because it produces what are arguably the most bizarre of the effects that arise from the special theory. However, Einstein also found by taking the same approach with Newton's laws of motion that there were two other effects produced by movement. As seen from the Earth, while our spaceship moves it will also be shorter in the direction of motion and will have become more massive. The faster it goes, the more this squashing up and increase in mass occurs. As far as the astronaut is concerned, it's the universe around the spaceship that squashes up in the direction of travel, shortening the distances that it has to fly.

The shortening of length in the distance of travel (in the observer's frame of reference) can be demonstrated with a light clock experiment where the clock is oriented in the direction of travel, rather than at 90 degrees to it. The maths is a little messier than the version in the Appendix, but the contraction emerges the same way. As for the change in mass, the easiest way to see it is from the equation that Einstein derived from the theory of relativity, $E = mc^2$, which we will explore in more detail on page 132. In the ship's frame of reference, the ship isn't moving, so has no kinetic energy. But for the observer on the Earth, the ship is moving, so in this frame of reference, the ship has more energy – and an increase in energy also means an increase in mass.

That increase in mass in anything moving is one of the reasons that the speed of light is seen as a barrier emerging from relativity. An object gets more massive as it moves faster and faster, meaning that it takes more energy to accelerate it. As it comes close to the speed of light, the energy required heads off towards infinity. However much energy is pumped into the system, it still isn't enough to get it past that light-speed barrier.

The strange effects of special relativity on length have no better illustration than a thought experiment that could be called the magic barn. This is a way to recreate the aspect of the TARDIS that so impresses visitors on *Doctor Who* – the barn appears to be bigger on the inside than it is on the outside – though only momentarily. In the early days of aircraft, showmen would fly their planes through the open doors of large barns – hence the term barnstorming. In this experiment, rather more oddly, it is a ladder that does the barnstorming and demonstrates that the barn can be bigger inside than it is outside.

Our experiment takes place in a large barn with doors at each end of it. The barn is 10 metres long. We start with both doors open and fire a ladder at extremely high speed – close to the speed of light – straight through the barn. The ladder is 13 metres long, so clearly it should not fit inside the barn with both sets of doors closed. But remember, according to special relativity, a moving object contracts in the direction of movement. The ladder is going so fast that as far as an observer standing next to the barn is concerned, the ladder is only

5 metres long. So, as the ladder enters the barn, our observer (with inhuman, lightning-fast reflexes) is able to briefly close both doors, enclosing the entire length of the ladder inside the barn. There is no problem doing this, as from his viewpoint, the ladder is 5 metres shorter than the barn.

Before the ladder has time to hit the back door, our operator opens that door and the ladder flies on its way. So for a brief moment of time, our ladder, which is 13 metres long, fitted entirely into the 10-metre barn with both of the doors closed. This is quite remarkable, but an inevitable result of the effects of special relativity. What really twists the mind, though, is to take a look at the same event from the point of view of an observer moving alongside the ladder, matching its speed.

In this frame of reference, the ladder isn't moving. So it remains the full 13 metres long. Also, from the observer's viewpoint, the barn is moving. And so it is the barn that becomes shorter in the direction of travel. What this means is that somehow, in the reference frame of the moving observer, we manage to have a 13-metre-long ladder entirely enclosed inside a barn that is significantly shorter than 10 metres in length. This just shouldn't work. Rather than solve the problem immediately, we're going to come back to it in a couple of pages when we have a warning about the nature of real objects and relativity under our belts. That will require another thought experiment.

Sometimes called 'the deadly rivet', this is the kind of thought experiment that gives physicists a bad name. In this setup we have a beetle, tucked away at the bottom of a hole in a table top. The hole is 10 millimetres deep. Just when the beetle was getting comfortable, a rivet comes hurtling towards the hole at close to the speed of light. The rivet is perfectly aimed so that its pin will enter the hole. But the beetle happens to know that the rivet is just 8 millimetres long, so it can sit safely at the bottom of its hole. (We have to assume, of course, special materials that can withstand this kind of collision. In the real world, the rivet would pass straight through the table or vaporise.)

However, at the last moment, the beetle realises it hasn't taken relativity into account. Let's assume for convenience of numbers that the rivet is travelling at around 0.87 c – that is, 0.87 times the speed of light. (This might not seem very convenient, but it helps with the maths.) The beetle, panicking, remembers that relativity has some kind of effect on the length of moving objects. Thankfully, the result is that a moving object becomes shorter in the direction of travel from the point of view of a fixed observer. This means that in the beetle's frame of reference, the rivet is only 5 millimetres long. Half the length of the hole. But from the rivet's viewpoint, it is the hole that becomes shorter. So will the beetle be squashed or not?

The beetle is still puzzling over this when it gets squashed. The reason why this happens becomes obvious when you think of what is happening to the different parts of the rivet at the moment of impact. The head of the rivet comes into contact with the table and stops. However, the pin part of the rivet doesn't know this has happened. It keeps on moving at close to the speed of light, and will continue to do so until a ripple passes down the pin, pulling parts of it to a stop. This ripple would typically travel at around the speed of sound. Realistically, the 'stop' ripple will not have time to reach the end of the pin before the pin reaches the beetle. After some small-scale vibrating to and fro, the pin will settle down and we will have an 8-millimetre-long rivet in a 10-millimetre-long hole. With the remains of a beetle at the bottom.

What does simultaneous mean?

The way that relativity makes aspects like the flow of time, length and mass no longer fixed can take a while to absorb. However, none of these is quite as shocking in undermining our view of reality as is the relativity of simultaneity, which says that whether or not two events are simultaneous depends on the observer's point of view. In a sense, this should be obvious once it's accepted that movement alters the flow of time, so that it is seen differently from, say, the Earth and a spaceship. However, it's hard to avoid the feeling that events in the universe come in a certain sequence, and that this sequence should not be influenced by relativity. What, for that matter, is the impact on causality – the idea that one thing is caused by another? If it were possible to alter simultaneity so that event A, which causes event B, comes after event B, then confusion would arise.

This is one of the reasons that physicists are so suspicious of the concept of a time machine that can send information back in time. It has the potential to disrupt causality. Think of a simple example – a radio transceiver that can be switched off by remote control. Imagine that the device is switched on. I use the transceiver to send out the signal that switches it off. So the device is now off. If there's a technical problem with a transceiver receiving its own signal – which I assume is something that is best avoided – then we'll just have a repeater transmitter, which takes the signal from the transceiver and rebroadcasts it, to be received by the transceiver a fraction of a second later.

Now I introduce my time machine. It's no TARDIS. All it can do is to take a radio signal and send it back one second in time. So here we go. We use the transceiver to send out the 'switch off' signal. It is rebroadcast by the repeater, and that signal is sent back one second by the time machine. A second earlier, the transceiver picks up the signal and switches off. So now the transceiver was switched off at the point that the original signal was sent. So no signal can be sent. We end up in a frustrating causal loop. If the order of interrelated events can be switched in certain ways, then we get in a lot of trouble. The whole idea of causality falls apart.

So with this in mind, let's explore why special relativity messes around with simultaneity. Usually when it comes to a major piece of physics, the examples used by the original scientist are too obscure to use in a more general explanation, but here Einstein's own example in an early book on relativity still works well.

We start by imagining a long length of straight railway line. There is a storm raging and two lightning bolts strike the line, many kilometres apart. We're going to say that the lightning bolts hit the line simultaneously. But how can we establish that the events truly are simultaneous? It's not possible to be in two places at once and directly observe the bolts hitting the line. So let's imagine instead that we put an observer half-way between the two points that the lightning is going to strike.

If our observer sees both flashes at exactly the same time, they are simultaneous. These days, we could also envisage having video cameras at the two locations, but video doesn't get instantaneously from A to B, so we would still have to make sure that the transmission times from the strikes to the central location were the same. If anything, the video approach makes the experiment harder to run than simply watching for the lightning flashes.

Now it gets interesting. We didn't put the railway line there just to have a handy lump of metal for the lightning to strike. Let's put another observer onto a train that is moving very quickly along the line, from left to right as we look at it. She, too, is looking for the two lightning flashes and passes the track-side observer at the very moment the flashes occur, from the point of view of the fixed observer. The observer beside the track sees the flashes happen at the same time – but the observer on the train doesn't.

During the time that light has been travelling to the mid-point from the two flashes, the train will have moved from left to right. So the observer on the train will see the flash from the right-hand end of the track happen before the flash from the left-hand end. We use light rather than, say, cannonballs fired from each end of the track because cannonballs don't provide a consistent measure for the moving observer, as the cannonball coming from the left travels slower, while the one from the right travels faster. However, because light always goes at the same speed, however quickly you travel with respect to it, it remains a useful measure. And one that shows that the idea of two events being simultaneous has become a relative one, rather than absolute.

You may by now nearly have forgotten the barn and the ladder, but it is time to get back to them. In that thought experiment, we had a ladder flying through a barn at high speed. From the point of view of an observer standing by the barn, the ladder is shorter than the barn, so he can shut the doors at both ends of the barn simultaneously with the ladder inside. But from the point of view of an observer flying alongside the ladder at the same speed, the ladder is longer than the barn. How is it possible from her viewpoint to shut both doors at the same time?

The answer is that it isn't possible to do so – but that it isn't necessary to do so either. Remember that two events that are simultaneous when an observer is fixed with respect to the environment in which they occur will not be simultaneous if the observer is moving. As far as the person standing by the barn is concerned, both barn doors are shut at the same time, trapping the ladder inside the barn. But in the moving observer's frame of reference, what happens is that the back door is shut first, then opened as the front of the ladder approaches it. The front door is then shut after the back end of the ladder has entered the barn. The two doors are not shut at the same time, but rather in the sequence back then front. The ladder never needs to be all inside the barn. Relativity of simultaneity has ridden to the rescue.

The relativity of simultaneity is a shock to the system. The idea that two events that I *know* are simultaneous are not simultaneous for someone else, just because that other person is moving, takes some getting used to. But once the principle is grasped, we have moved the universe into a more fundamentally relative state. Specifically, there is no more universal 'now' – because depending on how two observers are moving, they will not agree on how different events are positioned with respect to their individual view of 'now'.

That equation

There was one last trick up the sleeve of special relativity, which would produce the most famous equation in science:

$$E = mc^2$$

This was added by Einstein as something of an afterthought to his original work, in a short paper entitled *Ist die Trägheit eines Körpers von seinem Energieinhalt abhängig?* (Does the Inertia of a Body Depend upon Its Energy Content?), which he submitted in September 1905. There is little more than a page of working in this classic piece of scientific thought.

In the paper, Einstein used relatively simple mathematics to establish that the kinetic energy of a body diminishes if it should emit some light. He calculated that the reduction would be $\frac{1}{2}(E/c^2)v^2$ where E was the energy of the light given off, c the speed of light and v the velocity of the moving body. Einstein already knew, as most of us were taught at school, that the reduction in kinetic energy was also given by $E = \frac{1}{2}mv^2$, making E/c^2 the equivalent of m, the mass. Einstein had shown that the mass, m, equals E/c^2, which it is trivial to rearrange as the more familiar $E = mc^2$, a formula that did not, however, appear in the paper.

At the time this was a matter of interesting theory and little else. But Einstein added, as an apparent afterthought at the end, that it was possible that this theory could be put to the test with bodies whose energy content was highly variable, such as the newly discovered radium salts. It should mean that if his theory was correct, radiation would carry inertia from one body to another – in effect, radiation would produce pressure if it hit something (see page 78).

In this paper, Einstein did not observe that if the energy in matter could be released it could produce a devastating force indeed. But it did not take long for the implications to be explored and to be made dramatically real through the atomic bombs of the Second World War.

◇◇◇◇◇

Einstein had extended the reach of relativity into the heart of stuff and movement. Only gravity remained untouched by a relativistic approach. But not for long.

6 Gravitation

We have already encountered three of the fundamental forces of the universe – now we need to add in a fourth. If I were writing this from a Newtonian perspective, we could have brought gravity in with the other three fundamental forces when thinking about stuff, as the nature of gravity is strongly intertwined with the nature of stuff, but given a twenty-first century viewpoint it seemed more appropriate to hold it back for a while, both because the basic physics of motion can be dealt with without gravity, and because it turned out that gravity would provide one of the major stepping-off points where relativity proved necessary to explain the mechanisms of the universe.

As we have seen, to Aristotle and other Ancient Greeks, gravity (alongside levity) was an absolute tendency, a natural reflection of the character of the elements. Copernicus made this tendency meaningless, by moving the centre of the universe away from the Earth. Yet still gravity continued to keep us in place. There was no sensible explanation available for this until Newton made it universal and relative at the same time. Universal, because his gravity was the same everywhere in the universe, but relative because it depended on an attraction between any two bodies with mass, an attraction that increased in strength relative to their masses and that shrank with the square of the distance between them.

However, we are getting ahead of ourselves. Gravity occupies a bizarrely inverted position in our awareness. It is by far the weakest of the four fundamental forces. What's more, although we wouldn't exist *in practice* without gravity (because there would be no Earth or Sun without it and even if they did exist, a gravity-free planet would present life with some unique challenges), it is the only one of the fundamental forces that we can exist without in principle. The strong and weak nuclear forces are essential for the existence of matter, while electromagnetism enables us to interact with other matter and light. Without electromagnetism we couldn't see things, touch them, eat them or sit on them, for instance. Take any one of those three forces away and it's instant doom. Yet as astronauts regularly demonstrate, in the right circumstances gravity is an optional extra.

There are some provisos to that statement. Plants don't grow so well without gravity, as the roots don't know where to go and tend to form substandard structures. An experiment on the International Space Station showed that bird eggs were unlikely to develop to hatching in negligible gravity. And the human physiology deteriorates without a gravitational pull, both as a result of muscle atrophy and internal organs lacking their usual placing. But these are merely requirements for fine-tuning. Life could exist, if not have been formed, on Earth without gravity, whereas it is impossible to imagine that it could do so without any of the other three fundamental forces.

Despite its relative insignificance to our continued existence, gravity is the most blatant of the fundamental forces in making us aware of its existence. We were happily unaware of the nuclear forces for most of the time human beings have been around, and though the impact of electromagnetism is clear, it wasn't obvious what was behind our ability to interact with stuff until the end of the nineteenth century. But gravity is there to remind you every time you drop something or fall over. It has even been suggested that the earliest known physics experiments are performed when, as babies, we repeatedly drop items. Although it appears this is primarily done to irritate the parents, apparently it is the baby making a first exploration of the impact of gravity.

Introducing gravitation

In our model universe, things have been working just fine without gravity. But though we can use electromagnetism to build structures, electromagnetism doesn't offer us the universal attractive force of gravity, which gives nature the mechanism to form large-scale structures like planets and stars, and their larger-still groupings into clusters and galaxies. The existence of gravity, then, is very obvious in its results. It may be the weakest force, but it is the showiest in terms of its immediate outcomes.

Though humans had an early appreciation of gravity being there, we were less quick to understand exactly what gravity was. Arguably, even the characterisation of gravity as a force is not necessarily the best way to look at it. Yet it was first studied as a force, and that is how it naturally springs to mind. And as the structure of the solar system became clearer, the scope of that force was on the rise. First gravity was thought of as simply the tendency for heavy things to get as close to the centre of the universe as they could, but Renaissance astronomy brought extra implications that were quite different.

With the Earth shifted from the centre of the universe by Copernicus to become just another planet circling the Sun, and with Kepler's insistence on elliptical orbits for the planets, by Galileo's day there seemed a second role for a force like gravity – as a mechanism to keep the planets in their orbits, freed from the mythical structures of the ancient heavenly spheres.

Galileo did plenty of work on gravity on a local scale, though it is very likely that he never performed the famous experiment attributed to him of dropping balls of different weights off the Leaning Tower of Pisa to discover whether they fell at the same rate. Such measurements would be hard to make – nearly impossible with the limited technology of the time – and the story seems to have been conjured up as a publicity measure in Galileo's old age. Instead he preferred to study the acceleration under gravity of pendulums and of balls as they rolled down inclined planes, where the impact of gravity was more under control and easier to measure.

When it came to countering Aristotle's assertion that heavier bodies should fall faster, because there was more of the material in them that desired to get to the centre of the universe, not only did Galileo not need those falling objects in Pisa, he didn't even need an experiment. Appropriately, he turned the Greeks' own preference for a thought experiment back on themselves. Galileo was certainly aware of the armchair nature of Greek science and once wryly commented:

> I greatly doubt that Aristotle ever tested by experiment whether it be true that two stones, one weighing ten times as much as the other, if allowed to fall at the same instant from a height of, say, 100 cubits, would so differ in speed that when the heavier reached the ground, the other would not have fallen more than 10 cubits.

The mental experiment that Galileo undertook was to imagine that he had two balls of significantly different weights, where the heavier one did indeed fall faster as Aristotle had predicted. Galileo imagined letting both balls fall together, with a short piece of string joining them. This, he argued, would result in the lighter ball falling faster, dragged on by its heavier companion, while the heavier ball would be slowed down by the lighter, resulting in an in-between speed. However, look at the new object comprising both balls linked by the string and you have something that is, in total, heavier than either original. So the whole linked object also ought to fall faster still than the heavier of the two separate balls. This was a logical contradiction that seemed to make the original assumption invalid.

Although Galileo did not do the specific experiment, his work on allowing balls of different weights to roll down slopes under the influence of gravity and measuring their rates of acceleration provided the same result. In the end it's a difficult experiment to do, especially if one object is truly light, as the impact of air resistance then becomes considerable. It took David Scott, on the Moon as part of the Apollo 15 mission, to produce a truly graphic demonstration. He dropped a hammer and a feather at the same time to compare their fall. Without air to get in the way and slow down the feather, the experiment beautifully demonstrated the truth of Galileo's assertion.

The apple man

It was, of course, Galileo's philosophical successor Isaac Newton who came up with a comprehensive enough picture of gravity to cope with both the motion of the Moon and planets and the progress of objects that fell, though even Newton would stress the limitations of his understanding as he did so. He detailed what the expected effects of gravity would be, but said that he refused to speculate about how that gravitational attraction might work.

Along the way, Newton would struggle with the distinction between the absolute and the relative. As we have already seen, he believed there was such a thing as absolute space and time, yet though he was certain that gravity applied universally, in practice it seemed to detach itself from anything but a relative frame. For Newton, this could not be – not so much on scientific as theological grounds.

Newton's story is too well known to tell in any detail, but one aspect of the story that is worth revisiting is the matter of the apple. The story of Newton being inspired to think about gravity by a falling apple is often taken as a total fabrication. And certainly there is no truth to the suggestion that an apple fell on his head as he sat under a tree, setting off a light-bulb moment. However, the story of *seeing* an apple fall came from Newton himself, via the relatively reliable source of a contemporary, the antiquarian William Stukeley.

It's true that Newton was an old man when he told the tale, and he could have made it all up. It was, after all, around 60 years after the event he described. But sometimes a good story is also true, at least in its essence. Here is Newton's account, according to Stukeley, recalled from a visit to Newton's lodgings in Orbol's Buildings in London on 15 April 1726:

> After dinner, the weather being warm, we went into the garden, and drank thea [*sic*] under the shade of some apple trees; only he and myself. Amidst other discourse, he told me, he was just in the same situation,

as when formerly, the notion of gravitation came into his mind. Why should that apple always descend per- pendicularly to the ground, thought he to himself; occasion'd by the fall of an apple, as he sat in a con- templative mood.

This much seems fair. Why should Newton not have been inspired by that plummeting fruit to consider why it should behave the way it did? It is surely the mark of a good scientist to see something ordinary and think the extraordinary, asking 'Why?' when others are more inclined to take what has happened for granted. I do admit, though, that Newton's next observation probably required more thought than was likely to happen as he sat in the shade of the trees. Having said that, we are, after all, dealing with Isaac Newton who – despite his many limitations as a human being – was certainly a genius as a scientist:

> Why should it not go sideways, or upwards? But con- stantly to the earths center? Assuredly the reason is, that the earth draws it. There must be a drawing power in matter. The sum of the drawing power in the mat- ter of the earth must be in the earths center, not in any side of the earth. Therefore does this apple fall perpendicularly, or towards the center. If matter thus draws matter; it must be in proportion of its quantity. Therefore the apple draws the earth, as well as the earth draws the apple.

Once he had that point of view in place, including the essential relativistic component to the gravitational force that both the Earth pulls the apple and the apple pulls the Earth, Newton was ready to make the leap that gave this new theory a universal reach, taking in the heavens. According to Stukeley's description:

> That there is a power like that we here call gravity which extends itself through the universe and thus by degrees, he began to apply this property of gravitation to the motion of the earth, and of the heavenly bodys: to consider their distances, their magnitudes, their periodical revolutions: to find out, that this property, conjointly with a progressive motion impressed on them in the beginning, perfectly solv'd their circular courses; kept the planets from falling upon one another, or dropping all together into one center.

A second aspect of the story – the specific tree involved – is a little more shaky in its connection. There is an apple tree of the Flower of Kent variety at Newton's old family home, Woolsthorpe Manor in Lincolnshire, that dates back to the right period and could have been the actual tree. But then, Newton did say he was 'sitting under the shade of some apple trees', so even if the tree has survived from Newton's day, it might not be the specific tree he was looking at. It's certainly enough for most of the many visitors who still turn up to see that famous location. Whether or not the tree and its fruit played a role in his inspiration, Newton's place in the history of our understanding of gravity would become clear in the publication of his masterwork, the *Principia*.

Gravity and orbits

Building on work by Kepler and Huygens, plus some thoughts from his arch rival in the Royal Society, Robert Hooke, Newton pulled together the simple behaviours that Galileo had studied on the table top with the apparently unconnected movement of the planets in the heavens. Before publishing, having received a slight from Hooke, Newton expunged almost all reference to his contemporary from the *Principia*, though we do know from the letters they exchanged that Hooke presented Newton with at least one piece of the puzzle – understanding what is happening when one body orbits another.

Think of an orbiting satellite such as the International Space Station (ISS), as it travels around the Earth. (Technically the satellite and the Earth orbit each other, travelling around their combined centre of mass – the point where there is equal mass in each direction. But as the ISS is much smaller than the Earth, this is almost at the centre of the Earth. A system with more similar partners like the Earth and the Moon has a more displaced centre, so the Earth has a noticeable wobble as a result of the combined orbit. However, the Earth is sufficiently more massive than the Moon that the centre they rotate around is still inside the Earth, though displaced about three quarters of the way from the centre.)

It's easy to think of an orbit being a bit like holding an orange at arm's length and spinning around. If you stopped the orange's forward motion by ceasing to spin, it would still stay in the same place at the end of your arm. But the reality is quite different for the ISS. The force it feels due to gravity is straight towards the centre of the Earth. As a result of this force, the space station is accelerating towards the Earth. The ISS is falling. And it's only because the space station is falling that the astronauts feel pretty much zero gravity. They are sufficiently close to the Earth that they would otherwise feel around 90 per cent of the gravitational pull we feel on the surface.

If the ISS were not moving forwards, then, it would

plummet. But an orbit has a second component. The ISS is also moving at right angles to the direction in which it is falling towards the Earth. If we could magically switch off the Earth's gravitational pull, the ISS would fly off in a straight line at a tangent to the Earth's surface. So the combination of these two motions means that despite the ISS falling, it moves forward at just the right speed to keep missing. This is why, for any particular height above the surface, there is a specific velocity that is required for a satellite to stay in orbit.

This combined effect was the insight into the nature of orbits that Hooke gave to Newton. It provides a mental separation of the orbit into two easy pieces – a straightforward, straight-line movement, and a fall that involves acceleration due to the force of gravity between the two bodies. And Newton made it clear in his letter to Hooke that he, Newton, had never heard of this hypothesis before. This is not saying he was unable to come up with the concept on his own. He probably would have done so. But he does, in an exceedingly rare moment of generosity, acknowledge Hooke's contribution.

Hooke also argued that Newton stole a more mathematical part of his work in the assertion that gravity was an inverse square law. This is the physicist-speak way of saying that the force of gravitational attraction drops off with the square of the distance between two bodies that are attracting each other. Double the distance between them and the gravitational pull drops to a quarter of its value. Hooke had indeed mentioned that this ought to be the case, but when asked to prove it, in exchange for a considerable prize of 40 shillings (the equivalent of around £4,000 now in relative wages) offered by architect and fellow of the Royal Society Christopher Wren, Hooke blustered and never produced any kind of proof.

In reality, the idea that an inverse square law was involved was already in the air before either Newton or Hooke got their teeth into the problem. It seems to have been introduced first by French astronomer Ismael Boulliau. This priest and librarian was a foreign member of the Royal Society and wrote a book in 1645, 42 years before Newton finished the *Principia*, in which Boulliau argued that an inverse square law force would be required to make Kepler's explanation of his elliptical orbits possible (though Boulliau himself doubted that there was such a force). However, showing that the inverse square law applied mathematically was entirely Newton's work.

Quantifying attraction

It is difficult to get a clear picture of how Newton went about his original work on gravity by reading the *Principia*. In part this is because the book is deliberately obscure to keep its audience to the intelligentsia, and also because he reverted to geometry for much of the time, even though his core work depended on the much clearer approach of calculus. The outcome was an implied equation that never appears in the book, but that would become known as Newton's gravitational equation:

$$F = Gm_1m_2 / r^2$$

This tells us that the force of gravity (F) acting between two objects is proportional to the inverse of the square of the distance between the objects (r) – as the distance gets bigger, the force gets weaker, and this happens quicker and quicker as you get further away. It also depends on the mass of each of the two objects (m_1 and m_2). To complete the formula, we have G, which is a constant value that allows us to work out any one of the other items, given the rest of them. The value for G is not derived from anything else; it is assumed to be a universal constant that simply emerges from nature.

Those with a mathematical frame of mind might have spotted the danger in this equation – if r is very small, the force rises meteorically (to use a suitably gravitational adverb). As r comes closer and closer to zero, so the force heads precipitously off to infinity. This potential problem would resurface as an interesting issue when black holes cropped up. But for the moment, the obvious implication was that the r measurement clearly couldn't be the distance between the exteriors of the two bodies. For example, if m_1 were the Earth and m_2 a person, with r being the distance to the surface of the Earth, that person would be squashed flat by an infinite force. However, Newton managed to show that by dividing up a spherical body like the Earth into small chunks and considering the effect of each of the parts, gravity acted as if the entire mass were concentrated at the centre of the body. This meant that r must be the distance between the centres of the two bodies involved.

One of the ways that Newton tested out his theory was to undertake a thought experiment using the Moon. He began by coming up with a best guess, given the information available at the time, for the distance to the Moon. Then he imagined that the Moon had been stopped in its orbit and fell towards the Earth. Working out the acceleration, he extrapolated until his falling Moon had just reached the Earth's surface. At this point, an instant before catastrophic collision, he worked out that the acceleration the Moon experienced was the same as 'a Pendulum beating seconds in the latitude of Paris ... as Huygens observed'.

Newton had shown that the force that keeps the Moon in orbit is the same as the force that we experience when we drop something on Earth. As he put it: 'And therefore that force by which the moon is kept in its orbit, in descending from the moon's orbit to the surface of the earth, comes out equal to the force of gravity here on earth, and so … is that very force that we call gravity.' He also made a similar argument by comparing the effect of gravity and the action of an orbit for a small imagined moon orbiting so close to the Earth that it almost touched the mountain tops.

There is a satisfying example of the interlinked nature of physical laws that Newton's (unstated) formula tells us to expect something that Galileo had already discovered, namely that two bodies that have different masses fall at the same rate, or to be more precise, accelerate at the same rate under the force of gravity. We'll take a moment to explore this: if the thought of a few equations turns you off, feel free to jump forward to 'This is a good example …' below – but it's all less scary than a glance might suggest.

From Newton's second law we know the extremely useful formula:

$$F = ma$$

… linking the force (F) we apply to a body to the acceleration it experiences (a). Now it's pretty easy to see what is the acceleration that balls of different mass will experience. If we start with Newton's formula for the force of gravity, and label the two masses m_{Earth} and m_{Ball}, we have:

$$F = Gm_{Earth}m_{Ball} / r^2$$

Combining the two equations to get the acceleration on our ball, we get:

$$m_{Ball}a = Gm_{Earth}m_{Ball} / r^2$$

So the acceleration is just

$$a = Gm_{Earth} / r^2$$

The mass of the ball has disappeared. It doesn't make any difference what mass the ball has. Looking at what's left, we've got G, a constant, and m_{Earth}, which isn't going to change substantially. So as long as r, our distance from the centre of the Earth, remains pretty much constant, then the acceleration that a ball (or anything else for that matter) experiences on the surface of the Earth stays the same – as it happens, the value is around 9.81 metres per second per second.

This is a good example of how a physical situation that has relative components – the basic equation depends on the masses of both the Earth and the ball, and the difference separating them – can appear absolute because various factors are hidden. This is as a result of one aspect of the situation not changing (the Earth's mass), another appearing twice and cancelling out (the ball's mass) and a third that has a roughly fixed value in most circumstances we experience (the distance to the centre of the Earth).

An occult force

Once Newton's work had sunk in, it was clear that there had been a revolution in our understanding of gravity. Yet there was a gaping hole in the *Principia*, one that Newton himself acknow-ledged. At this stage he made no attempt to explain how gravity worked. Newton says in the *Principia*, 'hypotheses non fingo', which is often translated as 'I frame no hypotheses'. This sounds a little neutral for a phrase that was probably closer in intent to 'I'm not fudging things with hypotheses'. Newton notes that, unlike normal mechanical forces, gravity acts throughout the body, rather than just at the surface, but he resists the temptation to guess at a mechanism.

What we are left with is gravity behaving as an 'action at a distance'. Most of the time, what seems to be action at a distance proves to be an illusion. For instance, when I hear someone speak from the other side of the room, it might seem at first that my ear is responding directly to her remote voice. But we know that, in practice, what is happening is that her vocal chords are setting nearby air molecules in motion, that motion is then translated from molecule to molecule through the intervening air, and finally the nearest air molecules crash into my eardrum, producing the sensation of hearing. What appears to be action at a distance turns out to be a repeated local action crossing the medium between us and each stage depending on the direct contact of air molecules.

However, there seemed to Newton to be no such oppor-tunity for gravity to pass its influence from place to place by an intervening medium (others would attempt explanations based on this concept, as we'll see in a moment). And so many were left to interpret Newton's picture of gravitational attraction as something mysterious that could make things work remotely with nothing at all intervening. A term that was often used to describe it was 'occult', not in the magic sense that the word implies today, but rather to mean hidden. Even so, the implication of the term was certainly derogatory. Even the word 'attraction', which Newton had used to describe the action of gravity, proved to be a problem.

To us, attraction seems a perfectly reasonable description of the effect that two massive bodies have on each other due to gravity, just as we would use it to describe the effect of a magnet on metal (another action at a distance that was treated with suspicion at the time). It's now the natural word to use. Back then, though, the only common English usage of 'attraction' was in terms of having feelings for another person. To say that there was an attraction between the Moon and the Earth suggested that they fancied one another – which was certainly not Newton's intention.

It must have been very frustrating for Newton to be mocked, as he was, for this lack of a mechanism for his gravitational theory. We are now used to scientists coming up with mathematical models of nature that describe well how nature behaves without there being any real clarity on why that model happens to fit what is observed. If the model makes useful predictions and is constructed logically in terms of what it is modelling, then we are happy to make use of those predictions. Newton's universal gravitation made excellent predictions for the interaction of bodies with mass, so much so that it was all that was needed to successfully guide Apollo 11 to its successful Moon landing in 1969. But when Newton was working there was far more of an expectation that science should explain *why* things happened the way they did.

This emphasis on the 'why' was a hangover from the Greeks. Their theories had primarily been derived from assumptions about why things happened. The idea that earth- and water-based objects had a natural tendency to head for the centre of the universe involved explaining what was observed in terms of the *nature* of matter. Heavy matter needed to do this, and so that was what happened, just as it was the nature of a dog to chase a cat. Newton, and before him Galileo, had confined himself primarily to describing what actually did happen and finding mathematical descriptions or models of this that could be used to make predictions in other cases.

This is a fundamental shift that is at the heart of the thesis of this book. The Ancient Greek approach was a universalist one. Earth and water had a certain absolute nature and that would inevitably make for a particular outcome, unless something else interfered. Newton's gravity, while still having a universal behaviour for objects with mass, did not derive its predictions from the need for objects to take a particular action but rather quantified the force involved, arriving at it from the relative position and mass of the two objects. Newton's method made predictions by relating a mathematical model, at its simplest the equation on page 145, to the physical effects of gravity. This approach did not concern itself with the absoluteness of *why* this was happening, but rather focused on the relativistic measure of *what* would happen as a result of a particular set of circumstances.

To get a feel for the negative reactions Newton received because of a lack of understanding of his circumstance-based relativistic view, you only have to take a look at the response of two of his greatest contemporaries, both serious thinkers, yet neither of whom initially managed to grasp what was involved. The Dutch scientist Christiaan Huygens, who was often a supporter of Newton, dismissed the 'theories [Newton] builds upon his Principle of Attraction, which to me seems to be absurd'. At the same time, Newton's rival as the co-developer of calculus, the mathematician Wilhelm Leibniz, missed the significance of this change of viewpoint, calling it a 'return to occult quantities and, even worse, to inexplicable ones'.

The machine of nature

Although Newton had introduced one relativistic aspect to his work, he would never have accepted a fully relativistic view of the universe, because his understanding was so powerfully coloured by his religious views. Newton was not a conventional Christian – his beliefs were decidedly non-conformist at a time when this was still looked on with suspicion, and it was only by a certain amount of fudging that he got his position at Cambridge, where it was generally expected that fellows should be Anglican churchmen. But, without doubt, religion played a huge part in his life. It's notable that his great library, containing around 2,100 books, making it more than half the size of that of his Cambridge college, contained over four times as many books on theology as it did on physics and astronomy combined. (It also had six books on medals.) All the evidence is that for most of his life he spent far more time concentrating on theology and his other passion, alchemy, than he did on physics.

There was a real concern among his contemporaries that Newton's work somehow took God out of the picture, a position with which he would never have agreed. This potential outcome was clear from the writing of Newton's greatest academic fan, the French natural philosopher Pierre-Simon Laplace, whose work spanned the end of the eighteenth and the beginning of the nineteenth centuries. Based on Newton's work, Laplace imagined a purely mechanical universe in which, with enough information and mental power, the entire future of the universe, moment by moment, could be predicted.

Laplace made it clear that there was no place for a God in his view of reality.

One way that Newton managed to restore a role for God was in keeping the universe stable. It was pointed out that an unfortunate side-effect of universal gravitation was that any finite universe should collapse. Think, for instance, of a toy universe that is a finite sphere with stars evenly distributed throughout it. A star right at one edge of the universe will have no other stars pulling it towards the edge, but lots of stars pulling it away from the edge. Over time, all the stars would be pulled towards each other, ending up in a pile in the centre of the universe. This wasn't a good model for a perfect deity's creation.

To counter this problem, Newton first assumed that the universe had to be infinite. That way, any star would have stars pulling it in all directions. There would be no edge. (As we saw on page 37, it is possible to have a finite universe with no edge, but this did not occur to Newton.) The infinite universe Newton suggested had the potential to be stable. But in practice, there was still a problem. Even the slightest displacement from its proper position of a single star would be enough, over time, to start a collapse happening. Once it was out of place, it would feel more pull in one direction than another. There would be a local collapse that would spread across the universe like a chain of dominoes. Although the infinite scale meant there would always be some parts of the universe that survived, much of it would end up a mess, which once more was unacceptable. So after the initial creation of either a finite or infinite universe, Newton gave God the task of poking the stars back into position as and when they drifted.

Even though his theory of gravity had relative position and mass at its heart, Newton would have struggled to see the world through relativistic eyes, because his need for God to have an essential and practical role gave a fixed framework for all eternity. Humans may see things in a relative fashion, but God was the ultimate absolute. It's interesting that the introductory ode that opens the *Principia* (something we don't see enough in modern scientific papers), written by avowed atheist Edmond Halley who had funded the publication of the work, gives God a position. Even so, Halley uses the name of the Roman god Jupiter (Jove), and though this might be seen as a poetic reference to the true God by Newton, it's quite possible that Halley had a different idea. He certainly set his creator god some limits, voluntary or otherwise, writing:

> Behold the pattern of the heavens, and the balances
>   of the divine structure;
> Behold Jove's calculations and the laws
> That the creator of all things, while he was setting
>   the beginnings of the world, would not violate;

Whatever Newton believed God's role to be – and he added a short section to the end of the *Principia* where he emphasised that the wonder of nature should be taken to imply the existence of God – he did not believe that the deity was in favour of action at a distance. The attacks Newton received from the likes of Leibniz and Huygens must have stung, particularly because Newton did not believe in action at a distance either. Like many of his contemporaries, he thought that there must be some kind of material in space, an 'aetherial medium' or ether that played the same role in transmitting the attractive force of gravity as the air did in carrying sound to the ear. This ether was harder to detect than the effects of air, but Newton believed it surely must be there.

The difficulty was that it's much easier to use an intervening medium to send a push than it is to transmit a pull. Although various scientists and mathematicians would play with concepts like vortices in the ether that could provide that 'pull' effect, none was particularly satisfactory. As time went on and still no way of detecting a mechanism for gravity to work had been discovered, the ether was replaced by another mechanical approach that Newton had also considered as a possibility, though in a cruder form. This involved a universal bombardment of invisible particles, and variants of this particle shower theory would remain popular all the way up to the end of the nineteenth century, when William Thomson, Lord Kelvin, would be one of its last physicist supporters.

It goes something like this. Imagine that the universe is full of flows of invisible particles that are able to put pressure on the bodies they encounter. These particles act on massive bodies, but do not act on each other. (If this seems extremely unlikely now we know it is unnecessary, bear in mind that this is not dissimilar to the quantum theory of light, which acts on matter, but not on itself.) Now think what would happen to the flow of these particles coming, for instance, towards the Moon. From most directions, the particles would be coming in the same quantities and would cancel each other out, producing no net effect. But from the direction of the Earth, the Moon is in the Earth's particle shadow. The Moon will receive fewer particles from that direction. The result would be that the Moon feels a force that pushes it towards the Earth.

There has to be more to the theory than the simple description I have given, or gravitational pull would depend on the size of a body rather than its mass, and plenty of explanations for this variation were provided. But the theory does produce an inverse square law and, as such, it's not a bad start for an attempt to provide an explanatory mechanism for the attractive force of gravity. As we've seen, variants of this theory lasted for over 200 years until a young, rebellious German scientist changed for ever how we would think about gravity.

A happy thought

We've already come across Albert Einstein with his development of the special theory of relativity – a theory that is necessary to understand moving bodies once we know about the nature of light. That would have been enough for many a scientist to lay their claim to fame. And Einstein had also contributed in a big way to the foundations of quantum theory, the work for which he received the Nobel Prize. But his masterwork was yet to come in his transformation of Newton's work on gravity. Einstein's new way of looking at gravity, at a stroke, took away the problem of action at a distance.

Pragmatically, in terms of the numbers, Newton's work was doing a fine job. It delivered the correct results in all but a few minor cases. Yet it remained problematic in delivering a mechanism, and no one could find a way to iron out those minor cases, notably in predicting the orbit of Mercury, which did not behave quite as Newton's mathematics suggested it should. The starting point for Einstein, though, was not studying the *Principia* or any other work on gravity, but rather the result of a random thought that came to him as he sat at his desk in the Swiss patent office.

In *Einstein's Masterwork*, his recent book on the general theory of relativity, science writer John Gribbin strongly supported the suggestion that, while the special theory of relativity would have been developed anyway within a year or two of Einstein's paper in 1905, as several other physicists were working on the same track, the general theory – Einstein's assault on gravity – was a one-off, significantly ahead of anyone else, and might well have taken decades to reproduce.

According to Einstein, there was a clear moment when his thoughts began to firm up on gravity: in 1907, when he had what he called the happiest thought of his life. He later recalled: 'I was sitting in a chair in the patent office at Bern when all of a sudden a thought occurred to me: "If a person falls freely he will not feel his own weight." I was startled. The simple thought made a deep impression on me. It impelled me toward a theory of gravitation.'

With this simple thought, Einstein had moved gravity from a partially relative concept to true relativity. As we have seen, Newton's version of relativity did not provide a force of universal magnitude – it varied with the masses involved and the distance between them. However, given those aspects its force was universal. Yet Einstein realised the effects of gravitation, just like time and space, were not absolute but depended on the observer's frame of reference. It was different from special relativity, because that dealt only with frames in steady motion. For gravity to be explained required acceleration to be brought into the mix.

This 'thought' that Einstein had, sitting in his patent office chair, is now given the more impressive sounding title of 'the principle of equivalence', and it probably needs a little unravelling to see why it is so significant. Let's take a closer look at what is happening to a person who is falling freely. All this means is that he is falling under the influence of gravity, with nothing preventing his fall. So he is accelerating at the rate dictated by the force of gravity. And somehow that acceleration totally cancels out his weight. If we remove the impact of the air rushing past him, the falling person would be floating in space.

In fact, we've already seen this described, not for someone falling off a building but for someone in a circumstance that Einstein would not have envisaged back in 1907. It happens to the astronauts on the International Space Station. They are falling towards the Earth, and because of this they are weightless – they do not feel the effect of gravity. Experiencing the right level of acceleration cancels out gravity. But the orbiting motion means that they don't have the downside faced by the falling man of eventually hitting the ground.

The equivalence principle is remarkable because it gives us a powerful insight into the way things behave when accelerating that can run counter to common sense, but that is borne out experimentally. A great example is the unnerving balloon in the car. In this experiment, which can be carried out for real, a helium balloon is tied up in the middle of a car so it floats in the middle of the passenger compartment without touching anything other than its string. The car now accelerates forward. What happens to the balloon?

Our natural inclination is to think that the balloon will move backwards as the car is accelerating forwards. But let's check this using the principle of equivalence. We know that when something accelerates forward we feel a force pushing us towards the back. And according to the equivalence principle, this is the same as a gravitational pull in the backward direction. So our balloon is experiencing a gravitational pull towards the back of the car. What does a helium balloon do when faced with a gravitational pull? Because it's lighter than air, it moves in the opposite direction to the pull of gravity. So due to the gravitational pull towards the back of the car, when the car accelerates away, the balloon will float towards the front – the opposite of the common sense solution.

The leap that Einstein made with regard to the equivalence principle was to go from the obvious part of the thought that falling 'cancels out' gravity to the suggestion that the acceleration and gravity are entirely equivalent and indistinguishable in their impact. They are, in effect, the same thing. Once Einstein had taken on this astonishing viewpoint, it was possible to create a gravitational equivalent of Galileo's moving ship on which it was not possible to tell from experiments on board whether or not the ship was moving. But here the relativity is more general, dealing with gravity and acceleration.

In Einstein's equivalent, we are inside a spaceship with no windows. We feel a force towards the back of the spaceship. What the principle of equivalence tells us is that we will have no way to distinguish between two apparently different causes for this force. It could be that our spaceship is stationary on the Earth, parked on its rear end, and what we are feeling is the force of gravity pulling us towards the back of the ship. Or it could be that the motors (which are sophisticated enough to make no sound or vibration) are switched on, and the ship is accelerating forward at a rate equivalent to 1 g – the acceleration produced by gravity on the surface of the Earth. The two have exactly the same effects for any experiment we can undertake inside the ship.

Admittedly there is a small get-out clause. As described, there is a way to tell the difference between the two scenarios, because in the case of the gravitational pull, there will be a tiny difference between the force felt at the front of the ship, which is further from the Earth, and the back, so that an experiment at the front would detect slightly less attraction. However, this isn't so much a flaw in the equivalence principle as in the design of the experiment. Provided the requirement is that the two are indistinguishable at any specific location on the ship there is not a problem.

With our ship set up, we can perform a key experiment that opens up the path to the general theory of relativity (if not providing the maths to model it). Let's start with the ship moving at a constant speed through space. We set up a laser, running from one side wall of the ship to the other, producing a bright spot on the wall. Now we switch on the engines and start accelerating steadily. Galilean relativity – and for that matter special relativity – doesn't deal with acceleration. If you think back to Galileo's closed boat, it's easy to detect acceleration. Think what it's like when you are on a plane accelerating down a runway. Even if there was no sound or vibration you would know you were accelerating from the way that you were pushed back into your seat.

So it's no surprise that we can detect the effect of the acceleration in our spaceship. Because the acceleration is applied to the ship, but not to the photons of light that are crossing the ship, the light beam will no longer travel in a straight line across the cabin. It will curve towards the back of the ship. The greater the acceleration, the greater the curvature. Admittedly, with any acceleration we can realistically apply, that curvature will be tiny, but modern instruments could detect the shift in the bright spot on the far wall. More to the point, this is a thought experiment. We can apply an impractically large acceleration and can see in our minds a distinct curvature of the light beam.

So far, so unsurprising. Until you throw in the principle of equivalence. According to Einstein, we shouldn't be able to tell whether we are accelerating or under the influence of gravity.

And if that's really true, a gravitational field should make a light beam bend in exactly the same fashion. Einstein not only realised that, but extended the thought further. Much further. He could have simply thought that there was an effect on light similar to that on an orbiting satellite, a special extension of gravitational theory that covered light, bearing in mind that special relativity had shown that light has a kind of mass due to its energy, even though it has no inherent mass as a particle. (In fact, Newtonian theory does predict a change in the path of light because of this, but the effect is smaller than Einstein's theory predicts.)

For that matter, Einstein could have decided that the equivalence principle didn't cover this particular situation. Instead, he took the leap and wondered what would be the implications if light was just going on doing its thing, travelling through space in a straight line, but with the added twist that space itself was being curved by the effect of gravity. What if the presence of matter, of mass, caused space to warp? And this was where the idea got really interesting. Because not only would such a warp in space explain what was happening to light, it would also explain how, for instance, the Moon orbited the Earth.

Space warps

Seen through Einstein's new viewpoint – naturally an entirely relativistic viewpoint – the Moon was not being pulled off its natural straight-line course by some occult force that could act at a distance. Instead, the Earth was warping the space that the Moon was moving through, so the Moon's straight-line motion was twisted around the Earth. Gravity warped space. Or, more accurately, given that special relativity had ensured that space and time were inseparable, mass made spacetime warp and that was what we describe as gravity.

This warp can be hard to visualise because we are trying to imagine a twist in three spatial dimensions and one of time all at once. In effect, we need an extra dimension to imagine the warp taking place in – and our three-dimensionally obsessed minds find this hard to cope with. It's possible to get a feel for this by imagining the two-dimensional equivalent, using the third dimension as the warp direction. Which brings us to the mixed blessing that is the rubber sheet.

By far the most common illustration of the impact of the warping of space by, say, the Earth is to imagine space as a flat rubber sheet, held firmly around its edges. This flat two-dimensional space is our reduced-dimensions model of real space. We represent a beam of light – or the natural straight route of the Moon in flight – as a straight line drawn on the surface of the sheet. Now we put a very heavy bowling ball onto the sheet near the line. The ball sinks into the sheet, producing a distortion in the rubber 'space'.

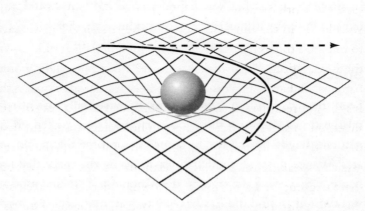

Figure 4: The rubber sheet and bowling ball model.

If we look at the line as it passes near the ball we will find that it is distorted by the warp. Instead of carrying straight on, it will now be curved, pulling the direction of the line towards the ball. The distortion of space by the mass has produced a curvature in the line that represents the beam of light or the path of the Moon. Of course, in the real world this process is more complicated because the curvature happens in three-dimensional space, not two.

Unfortunately, the trouble with models like this is that it is possible to stretch the analogy too far. And that very often happens when the rubber sheet model is used to illustrate the way that general relativity works. Because so far we have explained why the Moon orbits the Earth, but not why an apple falls from the tree. The apple isn't already moving, so it can't be just a warp in space that causes it to fall. The apple effect is usually illustrated on the rubber sheet model by imagining a small object sitting on the sheet. When the bowling ball is put in place, the object slides down the side of the indentation in the sheet caused by the ball. We have the falling apple, apparently attracted to the heavy ball.

Unfortunately, this explanation just doesn't work – it only seems to work because we are used to the way things are in our world. But in the universe model of the rubber sheet we have to question things more closely. *Why* does the object slide down the indentation? What makes it move? It's gravity. This would happen on Earth, but our rubber sheet model isn't on Earth. It's the entire universe. What would happen with a real rubber sheet out in space with no gravity? The object would just float there. It would have no inclination to slide down the sheet. (Strictly speaking the bowling ball shouldn't distort the rubber sheet without the Earth's gravity below it either, but we are considering a distorting effect to be a property of massive objects.)

The reason that we struggle with applying the rubber sheet

model to the apple is that it's natural for us to think of the warp in the sheet being a warp in space. But it's not. It's a warp in spacetime. *Time* is warped as well as space. And while the object starts off stationary in space, it isn't standing still in spacetime. So once more a warp can produce a change of motion. It can be helpful in imagining this to make use of a special diagram devised by Einstein's old maths lecturer, Hermann Minkowski.

Soon after Einstein came up with special relativity, Minkowski began to draw these diagrams to help understand the impact of relativity and the concept that Minkowski himself developed of spacetime. Einstein didn't really like them at first, probably because he didn't come up with them himself. But he came to accept them as a useful tool for envisaging the nature of spacetime.

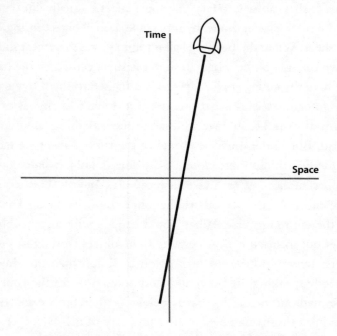

Figure 5: A Minkowski diagram of a moving spaceship.

In the simplest form of Minkowski diagram we have a graph with time running up the page and position in space running sideways. (Like the rubber sheet, we simplify the picture by taking out spatial dimensions. Here we display just one, although it is possible to have a second space dimension if the diagram is three-dimensional.) On a Minkowski diagram, a spaceship moving at a steady speed is shown as a straight line, plotting its position against time. This line, showing the location of the ship in spacetime, is known as its world line.

If we plot a Minkowski diagram of a static object – an apple, for instance, that we are about to release in the gravitational field of a massive object like the Earth, then before we release the apple, its world line is very simply a line going straight up the diagram.

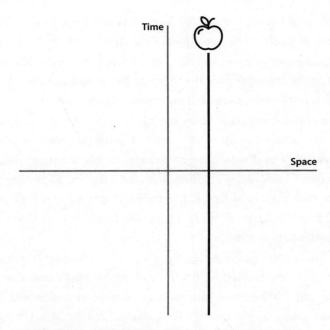

Figure 6: A Minkowski diagram of a static apple.

If we imagine the warp taking place in the Minkowski diagram for our apple, the apple's world line will be twisted away from travelling straight up the time axis into a curve across the space axis as well. The apple begins to move.

The geometry of curved space

Although Einstein thought through some of the implications of the principle of equivalence in 1907, he was primarily concerned with developing the early quantum theory for the next few years and it wasn't until 1911 that he came back to gravitation as his main focus, which it would remain until the triumphal publication of his general theory of relativity in 1915.

It might seem that a solid four years was a long time to add a spot of maths to an already existing theory – it certainly was a huge amount of time compared to his work on special relativity. But in the special theory, the mathematics is nothing that a good high school student couldn't cope with. To deal with the general theory, Einstein had to really stretch himself. The starting point was to move away from the kind of geometry we get taught at school. This Euclidean geometry is largely unchanged since Ancient Greek times and deals with flat surfaces, an assumption that is so ingrained that it mostly isn't noticed – it isn't even listed in the axioms or 'given assumptions' that precede the Euclidean theorems.

This mathematical obsession with flatness is quite surprising, because in the real world, flat surfaces are decidedly uncommon. It's a little different in our modern, manufactured world, but nature rarely comes up with flat surfaces, and this is never more so than on the surface of the Earth. We all know – and, for that matter, the Ancient Greeks knew perfectly well – that although a surface on the Earth looks flat when there are no significant hills or valleys, it really isn't, because we live on an object that approximates in shape to a sphere. And this three-dimensional shaping messes up Euclidean geometry.

One possible reason that the Greeks never spotted the big hole in their thinking is that Euclidean geometry took place in the world of the Platonic ideal – it operated in an imaginary perfect world where lines had no thickness and could be drawn perfectly straight, a world where all operations took place on an infinite plane. But as soon as that geometry is applied on the surface of the Earth, problems arise. For instance, an implication of one of the axioms of Euclidean geometry is that parallel lines never meet. It's what parallel lines *are* in the flat Euclidean world – lines that always stay alongside each other without ever coming together. But imagine drawing two parallel lines, each at 90 degrees to the equator on the Earth, both heading north. You'll see such lines on a globe; we call them lines of longitude. And they do meet – at the pole.

Similarly, on the surface of the Earth, the angles in a triangle add up to more than the 180 degrees total of the flat version. You can easily see that this is the case with the triangle formed by two lines of longitude, which have 180 degrees between them just from the two angles at the equator, plus the angle they make when meeting at the pole. Alternatively, on a different kind of curved surface that is concave, unlike the convex surface of the Earth, parallel lines diverge and the angles in a triangle add up to less than 180 degrees. It wasn't until the nineteenth century that the German mathematician Carl Friedrich Gauss dealt with this kind of warped two-dimensional space – but Einstein needed even more than Gauss had covered.

As it happened, in 1912 Einstein had been reunited with an old friend, Marcel Grossman, when Einstein took a post at the university where he studied as an undergraduate, the ETH in Switzerland. Grossman, who was already based there, recommended that Einstein, by now getting stuck on the mathematics required for his spacetime-warping gravity, explore the work of Bernhard Riemann, who had provided the definitive work at the time on the behaviour of multi-dimensional curved space. And this would eventually put Einstein on the right track.

Even so, the mathematics that Einstein discovered proved challenging for him, and he was in real danger of being scooped by the leading German mathematician of the day, David Hilbert. Berlin-based Hilbert had seen some of Einstein's preliminary work and set out to construct his own gravitational field equations, which he thought were ready to be published sooner than Einstein's. Luckily for Einstein, though, Hilbert made a last-minute error that meant there would be no dispute over priority. On 25 November 1915, Einstein submitted a paper entitled *The Field Equations of Gravitation*.

What Einstein had done was to replace Newton's equation

$$F = Gm_1m_2 / r^2$$

with something altogether more impressive – a collection of ten equations which, with the right simplification, would still produce Newton's results, but based on the concept of warped spacetime. Crucially, Einstein's new approach not only fitted with his vision of warped spacetime, it produced a range of testable predictions that were different from Newton's theory. Some of these were impossible to work on experimentally at the time, but one of them Einstein could make use of straight away.

As mentioned briefly above, the orbit of the closest planet to the Sun, Mercury, did not behave quite as Newton suggested it should. Its precession – the way that the orbit shifted with time – had a small difference from expectation. At the time there had been a suggestion that there was an unknown planet, given the name Vulcan, that was hidden behind the Sun and its gravitational field was influencing Mercury. But Einstein discovered to his delight that his new equations fitted precisely the observed behaviour of Mercury.

The equations of spacetime

There are a number of ways to write these gravitational equations. The simplest visually takes the surprisingly friendly form below:

$$G_{\mu\nu} + \Lambda g_{\mu\nu} = (8\pi G / c^4) \, T_{\mu\nu}$$

The apparently messiest bit of the equation in brackets is just a compound constant – the number 8, the familiar constant pi (π), Newton's gravitational constant G and the speed of light, represented as usual by c. The rest looks as if it is a trivial bit of algebra. But this is an example of the way that physicists can use notation where a single letter represents an entire equation, or, as is the case here, a whole collection of equations in a matrix. Each of the letters with subscripts (the Λ is another constant, which would cause Einstein a spot of bother) is a tensor, a mathematical structure that can have many forms, but in this case is a ten-dimensional mathematical object comprising a collection of differential equations – equations where the outcome varies with time and location. There is certainly a beautiful simplicity to this representation of the field equations, but beneath the surface, iceberg-like, is a massive collection of painful mathematical complexity.

One of the reasons for that complexity is that where Newton's simple formula dealt only with a mass-on-mass interaction, Einstein ended up taking in four contributions that would come together to describe gravity's spacetime-warping capability. As we have seen, the immediate implication of the principle of equivalence was that a moving object – or a beam of light – would find that its straight-line journey through space became a curve. This brings in the first contribution, as a total of six of the field equations, reflecting the three spatial dimensions and the potential to move in either direction through each. And we also need the warping of time as the second contribution, to make our apple fall. However, it's not just a matter of applying Newton's concept of the relation of mass to gravity in these different equations; as Einstein was to discover, to achieve his second contribution he would also have to bring in two other factors.

To begin with, there was the need to include energy. Special relativity had already shown the equivalence of mass and energy and the way that, for example, a moving body's mass would be greater from a location with respect to which the body is moving. Once $E = mc^2$ was on the scene this might well have seemed an obvious conclusion. So Einstein had to consider the gravitational effect of energy. But less immediately obvious to the non-physicist, Einstein also had to deal with pressure, which itself generates a small gravitational component.

These direct space and time effects provide the two factors in Einstein's second contribution. His other two contributions are smaller, but nonetheless have to be added in for exact calculation and in some circumstances can be important. One is frame-dragging. This is a gravitational effect where a massive rotating body pulls spacetime around with it in a kind of vortex. It's a bit like twisting a spoon very quickly in a jar of honey. As the spoon rotates, it will pull the nearest honey with it. This will then pull the honey a little further out, though somewhat less. Eventually a stream of honey will be rotating with the spoon.

Frame-dragging by the Earth has been observed by a number of experiments. Some have suggested that this effect is our best hope for a backward-travelling time machine, making use of the way it distorts spacetime – though other scientists query the validity of this analysis. Whether or not this is the case, the effect is present and occurs because special relativity requires that a moving massive body produces a small gravitational pull at right angles to its motion. It is this sideways pull that provides the frame-dragging when a body rotates.

To see where the pull comes from, it's easiest to think of a simple setup where a stationary object – a heavy ball, say – is sitting between two strips of material that have plenty of mass. The top strip is moving left to right, while the bottom strip is moving right to left at the same speed, in the object's frame of reference. Both strips have the same mass, so the object doesn't feel pulled towards either of them.

Figure 7: Strips and ball.

However, special relativity doesn't allow us to make a generalisation like 'both strips have the same mass' – something that is going to have an impact on gravitational effects. Imagine that we now fly past the experiment from left to right with the same velocity as the top strip. From our viewpoint, the top strip is not moving and the bottom strip is moving at twice the speed it was. We know from special relativity that a moving object has an increased mass. So from our frame of reference, the bottom strip has more mass than the top strip. This means that our ball should be pulled towards the bottom strip.

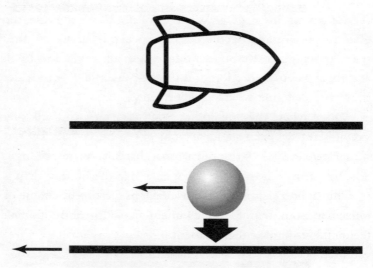

Figure 8: Strips, ball and spaceship.

This is in danger of producing an impossible outcome. It can't be that the ball doesn't move up or down if we are stationary but starts to move down if we are moving. That's taking relativity too far. So there's only one alternative. From the spaceship's frame of reference, the ball is moving from right to left, where it wasn't moving for the static observer. So that motion of the ball must produce a sideways gravitational pull that balances out the extra gravitational pull from the increased mass in the bottom strip, preventing the ball from moving.

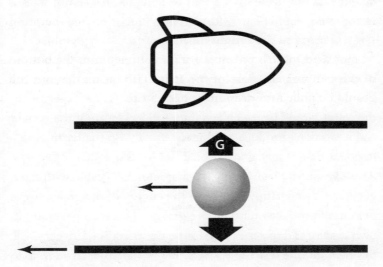

Figure 9: Strips, ball and spaceship with balanced forces.

This gravitational effect at 90 degrees to the direction of movement is sometimes called gravitomagnetism. This is a distinctly misleading term as there is no magnetism involved, but it was used because it has some similarity with the way that magnetism is produced at right angles to the direction of travel of a moving electrical charge. This effect results in frame-dragging for a rotating body and provides Einstein's third contribution for his equations.

With three contributions in place, the final part comes, rather bizarrely, from the gravitational pull of gravity itself. Gravity itself generates more gravity in a kind of feedback effect. The reason for this is obvious when you take in the real nature of one of the types of energy that you were probably taught about at school – potential energy. It takes energy to, for instance, take a heavy weight up the Leaning Tower of Pisa. You will have been taught that you were giving that weight *potential energy*. This is energy that would be transformed into kinetic energy – motion – if you dropped the weight off the tower.

But where is that potential energy coming from? It is a result of the gravitational field of the Earth. If you transported the Leaning Tower into deep space and did the same experiment, the result would be much less interesting. Let go of the weight and it wouldn't produce any kinetic energy. So the Earth's gravitational field – any gravitational field – is a source of energy. As we've already seen, energy, being interchangeable with mass, generates a gravitational pull. So the energy of the gravitational field itself produces a little more gravity. This then generates its own gravity and so on – but because the effect is relatively small, this is an infinite series like $1 + \frac{1}{2} + \frac{1}{4} + \frac{1}{8}$ ... which adds up to a finite sum and can be calculated and included in the equations.

Testing the theory

Not surprisingly, solving a set of equations like this with all these components (bearing in mind also that we don't just have one equation for space, but six) is non-trivial. There is still no universal solution for the full gravitational equations, but it proved relatively easy to produce a result for special cases, making it practical to use the general theory of relativity to make predictions that could be tested.

As we have seen, Einstein had already been able to test his theory against the existing observed variation in the orbit of

Mercury from the predictions of Newton's theory, and the outcome was an excellent match. But the new theory ideally needed experimental data found after the theory was put together to reinforce it. If, for instance, the bending of the path of light by matter could be demonstrated to match the theory's prediction, then it would be a triumph for Einstein.

They couldn't do my gun and laser experiment (see page 77) – not only did they lack an infinite Earth, but lasers wouldn't be invented until 1960. But a star like the Sun would warp passing light enough to be detected. To make this a usable test, though, the passing light from more distant stars needed to be visible near the Sun. Generally speaking this isn't possible, as the Sun's intense light washes out the starlight. We tend to think of the stars not being in the sky during the day, but of course they are, we just can't see them. However, there is one time when stars can be seen during the day, even close up to the Sun's perimeter: during a total solar eclipse, when the Moon blocks out the sunlight.

The first expedition to test the general theory of relativity came before Einstein had totally completed his struggle with the mathematics. (In most cases, observing an eclipse requires an expedition, as it is visible only from a small slice of the Earth's surface.) The German scientist Erwin Freundlich attempted to take measurements in an eclipse that was visible from Crimea in August 1914. But Freundlich got caught up in the repercussions of the outbreak of the First World War. He discovered it was not a good idea to be on enemy territory with large telescopes and was arrested as a spy. Similarly, the eclipses of 1916 and 1918 passed by with the confusion of war making scientific work impractical. And so it was on 29 May 1919 that two expeditions attempted to make the measurements that would ensure Einstein's lasting fame.

Both expeditions had problems with the weather and their technology, but after analysing the data, British astronomer Arthur Eddington was able to announce that the observers had confirmation of the displacement predicted by Einstein. Since 1919 there has been some doubt about whether this was really the case. Eddington's teams were working from a very small number of usable exposures and were right at the limits of the accuracy of their equipment to be able to distinguish the relatively small difference between Einstein's predictions and Newton's. Eddington may have indulged in some wishful thinking in his firm announcement of confirmation – but as it happens, he got it right, as subsequent, far more accurate, experiments have all confirmed Einstein's predictions.

Before we leave the equations of general relativity behind, it's worth mentioning that extra constant Λ, the Greek letter lambda, which turns up in the second term of the field equations:

$$G_{\mu\nu} + \Lambda g_{\mu\nu} = (8\pi G / c^4) \; T_{\mu\nu}$$

This is sometimes called the cosmological constant. Einstein famously added it because, without it, the equations seemed to predict that the universe should collapse. Einstein was of the opinion that the universe was in a balanced state with a constant size, so arbitrarily added this constant as a fudge factor to enable the universe to resist the contracting force.

He would later refer to this as his greatest mistake, though as it turned out, his error was not in putting the constant in place, merely in the value he selected. When it was later discovered that the universe is not of constant size, but is expanding, a different value for this constant ensured that Einstein's equations delivered the appropriate value. In effect, Λ represents the mysterious contribution of what is labelled 'dark energy'. What had been considered a mistake turned out to save an equation that otherwise would not quite match reality.

Making waves

The general theory of relativity has stood the test of time. Although we still have a big theoretical gap in trying to pull relativity into line with the other three forces of nature, Einstein's work has proved resilient to test after test. At the moment gravity resists a quantum form, making the borderline between the physics of the very small and the very large fuzzy. Much work has been put into theories such as string theory and quantum loop gravity, which provide a quantum version of gravity compatible with general relativity, but as yet none is testable or entirely satisfactory. But this doesn't stop the general theory producing elegant and effective predictions, notably that of gravity waves.

In a rather error-prone paper in 1916, corrected and clarified in 1918, Einstein predicted that just as a moving electrical charge produces an electromagnetic wave, a moving massive body should produce a gravitational wave. So robust is general relativity that there has not been much doubt that such waves exist, and they have been detected indirectly since the 1970s, when a binary pulsar was discovered to have a varying frequency.

A pulsar is an ultra-dense collapsed neutron star, rotating very quickly, giving off a lighthouse-like beam of radio waves. In 1974, the Arecibo radio telescope in Puerto Rico detected the pulses of a source that was changing in frequency. It was suggested that this was caused by the pulsar being in a binary system with another star. As the pair rotate they should be strongly generating gravity waves, carrying off energy, which would change the orbital frequency – and this is exactly what has been observed.

Such indirect observation is interesting, but in 2016 a much bigger breakthrough was made by LIGO (that's the Laser Interferometer Gravitational wave Observatory). The detectors picked up gravity waves from a distant event, waves that appear to be those that would be expected from the dramatic outburst resulting from the merger of a pair of black holes. This is not only valuable data in its own right, but has the potential for providing a whole new way to observe the universe. At the moment all our telescope work to probe the cosmos depends on light, from low-energy radio, through the visible, all the way up to X-rays and gamma rays. But light is always influenced by matter: scattered, absorbed, generally messed up by objects in the way. Nothing, however, stops gravity. It could even see past the circa 300,000-year barrier – the point in the lifetime of the early universe when it became transparent and light could pass through it.

Despite the huge and valid enthusiasm that followed the 2016 discovery, we do need to be a little wary of promises of a transformation of astronomy any time soon. This is because gravitational waves are very difficult to detect. Gravity is, after all, ludicrously weak compared with the other forces of nature, and in looking for gravity waves we need to detect tiny fluctuations – around 1/1,000,000,000,000,000,000,000,000th of the background level. This is far smaller, for instance, than the impact of a car driving past a gravity telescope. And unlike light, there is no way to screen out unwanted fluctuations.

Increasingly sensitive gravity wave detectors have been built for decades, but it wasn't until the enhanced version of LIGO detected the black hole collision in 2016 that any detection whatsoever had been made. LIGO is made up of two detectors, around 3,000 kilometres apart, in Hanford, Washington, and Livingston, Louisiana. Each consists of a pair of tubes around 4 kilometres long in an L shape. A laser is blasted down the vacuum-filled tubes tens of times before the beams are brought together to interfere with each other. If strong enough gravitational waves arrive, the tiny fluctuations they produce in spacetime should result in a subtle difference between the light going down the two arms, resulting in a shift in the interference patterns – and this is what happened.

But it's not enough simply to detect a shift. It might be caused by a passing vehicle, waves on a beach, distant earthquakes – anything that might slightly move the equipment. This is why the two well-separated devices are used. If a blip arrives at both simultaneously, it is unlikely to have a local cause (though even here the scientists have to eliminate, for instance, any earth tremor originating between the detectors).

Because of the potential for false sources, detection has to be confirmed in a probabilistic fashion. After all, it could just be a coincidence that a blip occurs at both detectors simultaneously. The scientists have to decide which levels of detection to accept and which to reject as false signals. One of the ways they do this is to shift one signal in time, seeing how often coincidences happen that definitely *aren't* caused by the same signal and so trying to ensure that their detection is valid. This means that, for the moment at least, gravity observatories will never have the clear, simple detection we expect of a light-based telescope.

The hope is that LIGO's work can be built on with a successor called eLISA, which gets around many of the problems of the current detectors by moving into space, away from the possible disturbances faced by an observatory on the ground. Instead of LIGO's 4-kilometre arms, eLISA (Evolved Laser Interferometer Space Antenna) would use 1-million-kilometre beams through space, making it far more sensitive. The eLISA project replaced the larger LISA, cancelled in 2011, and at the time of writing is planned for 2034, though a 'pathfinder' satellite has already been launched to test some of the technology involved. Gravity waves show great promise for a new type of astronomy – but we've a long way to go before they are everyday practical tools.

◇◇◇◇◇

With gravity in place, all the fundamentals of the physical world have been assembled. Our DIY universe has come close to the real thing. Yet look around the Earth and there's something else making a huge impact everywhere that isn't sufficiently described by any simple combination of space, time, stuff, movement and gravity. Something that has made the Earth remarkably different from the other rocky planets in our solar system. That something is life.

7 Life

◇◇

As we have built up our model it has become closer to the real thing. But from the point of view of establishing our place in the universe, there is a clear requirement to add life to the mix. There are three essentials here – getting life to come into existence at all, getting from simple single-celled life to complex single-celled life, and making the long journey from a complex single cell to a human being.

More than stuff?

I regularly give talks to primary school children in which I explain what science and science communication are really about. The bit that always goes down best is when I show how both science and its communication are often about taking what appears to be a collection of boring, everyday facts and uncovering a far more exciting and wonderful reality lying beneath. This is the total opposite of Keats' complaint in his poem *Lamia* that (natural) philosophy (i.e. science) spoiled things by 'unweaving the rainbow'.

Having asked how old the children in the audience are, I open up the concept of age by pointing out that parts of them

– their blood cells – are only days old, while the egg that they came from was formed when their mothers were born, so they can add their mother's age to theirs. And that's only the beginning, because the atoms in their bodies have previously been in plants and animals and other people. They've been in kings and queens and dinosaurs. Pretty well all the atoms in their bodies were already here when the Earth formed around 4.5 billion years ago. So that makes them at least 4.5 billion years old. But they're even older than that, in fact.

The heavier atoms, like the carbon in their flesh and the oxygen in the water in their bodies, were made inside stars which then exploded and spread the atoms across space to eventually become part of our solar system. They were made perhaps 7, 8, 9 billion years ago. And the lighter atoms of hydrogen that are in both flesh and water have been around since shortly after the universe began, 13.8 billion years ago. A lot of what they are, then, of what makes them up, dates back to the beginning of the universe.

This produces genuine shock and awe in the children (in a good way). They are fascinated by the lasting nature of the matter that makes them up, its near-absolute existence while everything else changes around it. And though, as we have seen, relativity is the ruling reality between different pieces of matter, as individual particles, inanimate matter continues to have this kind of absolute existence. But an absolute existence is certainly not the case when it comes to life. All the life we know of is highly dependent on its frame of reference, at the mercy of evolution.

Cue evolution

As we will discover, we are still largely in the dark over the origins of life, but the scientific community pretty much universally agrees that the development that followed was driven by evolutionary forces. And to understand evolution, a frame of reference is just as important as it is in describing the simultaneity of events in time. In our DIY universe we will require 'evolution by natural selection', describing the way that living organisms can accumulate changes from generation to generation which give some kind of benefit in their particular frame of reference (this is the natural selection part) − a frame that will include the environment, predators and competitors for food and sexual reproduction.

For the majority of the time for which we have records, the explanation of life has been dominated by absolutism. Until very recently in the history of humanity, the generally accepted view was that all life was created, in the same forms as it has now, soon after the beginning of the universe by a creator. We still acknowledge this indirectly when we use the word 'creature', with the implication that an organism has been created by someone or something. Of itself, evolution does not preclude a creator. Any doubt about that would be easy to show, given the vast variety that is found in one single species, the domestic dog. The differences between a Chihuahua and a Great Dane − both the same species − aren't the result of undirected evolution by natural selection, but are the directed outcome of very unnatural selection.

However, as long as we accept that evolution by natural selection is the key mechanism for getting from the first forms of life to the diversity we see today, we need to move to a relativistic view. The intellectual debris of the earlier creationist model often causes a misunderstanding about the nature of evolution, suggesting that the process is somehow driven to improve living things, making them ever more complex as if there were some kind of idealised goal (possibly human beings) towards which evolution was gradually pushing development. But the reality is nothing like this. There are, indeed, examples of evolutionary development where organisms have become less complex than their ancestors. Some large and complex bacteria-like cells, for instance, have proved to have lost some of their complexity and yet still survive, evolving into a simpler form that fitted well with their particular reference frame.

The reality of evolution by natural selection involves holding random changes up against the frame of reference that includes the environment, competitor species and competitor variants within the same species. If a change introduced by genetic modification or by environmental pressures on epigenetics (the biological mechanisms outside of the genes) gives a species a better fit in that frame of reference, then it is more likely to pass on its genes to its successors and more able to keep the species variant going. It is all about the frame of reference – an individual, isolated member of a species cannot evolve through natural selection in any meaningful way.

We will come back to evolution when we have begun to populate our world, but first we need to get hold of a basic form of life – and that can only be possible if we know what life is.

What is life?

Unfortunately, just as St Augustine knew perfectly well what time was, but struggled to be able to describe it to anyone who asked, we are totally comfortable that, say, a rock isn't alive where a daffodil or a dog is – but we have distinct difficulties pinning down just what being a living organism implies, not helped by borderline cases like viruses which have some of the characteristics of life but lack some of what are usually considered essentials.

At one time, it was felt that life required an extra 'something' that a dead or never living object did not have – a kind of built-in energy, often called the 'life force', which powered its existence. This seems a natural, common sense concept, because living entities can do things in a way that lifeless objects can't – and being able to do things generally requires energy. Like many concepts that feel like common sense but aren't supported by science, the life force lives on in pseudo-science, medical methodologies based on 'ancient wisdom', metaphor (ever been told you looked 'full of life'?) and fantasy novels, but there is no scientific evidence for the existence of some special, different kind of biological energy over and above the familiar types of energy like chemical, electrical and kinetic.

Lacking a clear description of what life is, biologists have resorted instead to listing the symptoms of life, what being alive entails – the typical occurrences and activities that are expected when life is present. You may remember learning seven 'life processes' at school. These are usually given as:

- Movement – even plants move over time: watch a sunflower follow the Sun
- Nutrition – consuming something, whether that something is plants, animals or sunlight to generate energy
- Respiration – technically the process by which energy is produced from the 'food' source, and often but not always involving oxygen
- Excretion – getting rid of waste matter
- Reproduction – making new copies of themselves (often with variation) to continue the species

- Sensing – having some interaction with what is around them, usually by detecting forms of energy
- Growth – though not a constant throughout life, all living things grow at some point in their development

There are significant problems with taking this approach, because to reach the full set of processes we need to consider an organism as a whole. By this definition, for example, *you* are alive, but the cells that make you up are not, because they exhibit some of these qualities but certainly not all of them. While it's possible that life is an 'emergent' quality that arises from the whole being more than the sum of its parts, it somehow seems wrong, as well as linguistically bizarre, that we should think that a living cell isn't alive. Ask a cell biologist and she will have no doubt that cells are alive, as Jenny Rohn makes clear:

> [That cells are alive] is never more evident than when you've had a bad day in the lab and you end up killing your cell cultures by mistake. Cells that are alive metabolize, and divide, and move around – if you film them with timelapse microscopy, they are amazingly dynamic, quivering and pulsating and sending out probing little fingers (filopodia) and feet (lamellipodia); some cells even crawl around. And of course, they reproduce themselves – some endlessly, like immortal cancer cell lines. When cells die, they retract all their fingers and feet, and round up – their nucleus disintegrates and they sort of explode. Then they are utterly motionless, never to rise again. So in my view, this is clearly the difference between life and death!

However, the individual cells of a complex organism like a human being can't live for long on their own without external support – so what they have may be life, but not as we usually know it. As we've already seen, another classic example of a dead/alive conundrum is the virus. Viruses have many of the characteristics of a simple living single-celled organism like a bacterium, but lack a true metabolism. Similarly, they do not contain the mechanism required to reproduce themselves, making use instead of a remarkable ability to hijack the reproductive mechanism of the living cells that they attack, using that to duplicate. Viruses exist in a strange twilight world of existence where they are not truly classified as living, yet are certainly not as lifeless as a piece of stone.

Perhaps all we can realistically do is to label as living anything that has appropriate mechanisms, direct or indirect, to fulfil the processes of life – allowing, for instance, for the indirect pathways provided by the interaction between cells to form a living whole. If that's the case, we probably should begin to treat viruses as living things. Either way, we have in life what is sometimes described as an epiphenomenon, an emergent property that goes beyond the capabilities of its components. We also know that life is something that isn't easy to just make happen, and yet we know that the Earth is positively teeming with it.

It's ... not alive

In the 1950s, it seemed as if the need for a creator had been done away with for ever, because it was thought that, with the right environmental conditions, life would be an inevitable outcome. Stanley Miller, a PhD student working under the eminent biologist Stanley Urey at the University of Chicago, set up an experiment which reproduced the conditions that were then thought to have existed on Earth when life began. In a reaction vessel containing water, methane, ammonia and hydrogen, Miller sent electrical discharges through his mixture to simulate the impact of the lightning bolts that were thought to have ravaged the early Earth. This provided the energy to get things started in good Frankenstein fashion.

After running the experiment for a week, then applying a 'killer solution' to ensure that there was no external contamination, Miller discovered that relatively complex organic molecules had formed themselves, notably discovering traces of the smallest of the amino acids, glycine. Such chemicals are often called the building blocks of life because they are the components from which life's key structures, proteins, are assembled. Remarkably, when sealed containers of the output from the experiment were re-examined in 2007 with more sophisticated tools, it was discovered that more than twenty amino acids had been produced on a very small scale. We now know that the seemingly complex forms of amino acids are assembled relatively easily – they have been found, for instance, naturally occurring in space.

However, all that Miller's experiment showed was how relatively simple organic chemicals like amino acids could be assembled from the compounds that were thought to be around at the time when life began on Earth. Unfortunately, the biologists involved made assumptions about conditions on the young planet which would prove to be unfounded. And there is a far greater leap from amino acids to life than anything that had been achieved in Miller's reaction vessel.

To think that you are close to life when you have amino acids is a bit like thinking that you have nearly constructed a modern car if you have a box full of gear wheels. Take DNA, for instance

– something we recognise as an essential component of almost all life on Earth. DNA provides the working instructions for the molecular machinery that will assemble amino acids to form proteins. The chromosomes found in living cells are individual DNA molecules, which thanks to DNA's unique structure act as a data store.

Each DNA molecule consists of a set of 'base pairs', complementary pairs of organic compounds called bases, which spell out the data in a code of four different molecules, cytosine, guanine, adenine and thymine. These pairs are held in place by the twin spirals of sugars that form the famous double helix. DNA is a very sophisticated molecule because it not only carries this genetic code, but is built in such a way that it can split in half, and each half can then be used to recreate the whole, essential to enable the DNA to be present in dividing cells as an organism grows. This clever trick is enabled because the base pairs always join up with the same partner, cytosine with guanine and adenine with thymine.

DNA might have a simple enough repeating structure, but the molecules of DNA that make up chromosomes can be immense. The biggest ones discovered so far have billions of base pairs, making it hard to see how simply having an organic soup that contained some amino acids and sloshing it around would somehow lead to the magical formation of DNA. Even in Miller's day it was recognised that DNA was unlikely to be produced this way, so the assumption was of an early 'RNA world' where life depended on the similar but simpler molecule RNA to carry the information required for reproduction and development. But there was not a plausible mechanism for RNA to form either.

Even if the soup had produced the relevant chemicals from

which to construct the information-bearing molecules of life, these are in some ways the simplest parts of a living cell. It's the sophisticated molecular machines that read off the messenger chemicals, construct proteins, and make use of the energy-carrying molecule ATP among other things that dwarf simple molecules in their complexity.

This term 'molecular machines' sounds like an exaggeration or a metaphor. But it is nothing of the sort. They may look messy to an eye that is used to the straight-line simplicity of designed engineering, but some of the mechanisms inside cells, necessary for life as we know it, rival sophisticated machinery in their complexity. They are literally mechanical devices operating in a nanoscale world, well beyond the capabilities of our current technology. And all the evidence is that this machinery was present inside the first common ancestor, from which *all* current life on Earth has evolved. A single ancestor that pretty much has to have existed because of the shared structures that the varying forms of life on our planet have in common.

As it happens, those assumptions made by Miller about the conditions on the early Earth were way out. It wasn't his fault – it was the best guess from Earth science of the time – but evidence that has been found since does not bear out this particular collection of compounds. The mix was chosen because it resembled the atmosphere of Jupiter, which was thought to reflect the make-up of our early atmosphere, assuming a shared heritage in the early formation of the planets. However, research based on zircons (see page 198) has shown that the atmosphere 4 billion years ago lacked the crucial methane and ammonia used to build those amino acids. Instead, it was primarily nitrogen, carbon dioxide and water vapour. And even if Miller's assumptions had been correct, all he had done was to push back the boundary a little – biologists of the time were no closer to discovering how it was possible to make the leap from a bunch of chemicals to the complex structures of even the most simple living cell.

For that matter, there is the problem of what provided the power source for the very first origins of life. While Miller had demonstrated that an electrical discharge could provide the energy to push forward some chemical reactions, using lightning that it was thought was present in unusually large quantities on the young Earth, no one was suggesting that the first life was itself plugged into the cosmic mains. It needed to sustain itself, not just be brought into being.

An essential for life is nutrition – a mechanism for taking energy from an external source and converting it into the chemical energy needed for the living organism to function. There is no living creature that can feed on the electrical discharges of lightning and there is no envisaged mechanism for one to do so, even if lightning had been present all the time, which it wasn't. If life had been created by repeated bolts of lightning then there had to be some other mechanism to keep that life functional.

Some wondered if the energy source could have been light, bearing in mind the high levels of ultraviolet (UV) that were present on the early Earth, unprotected by an ozone layer. But we know the problems we have today with the meagre amount of ultraviolet that gets through to us, causing cancers and genetic damage. Biochemist Nick Lane writes: 'UV is too destructive, even for the sophisticated life forms of today, as it breaks down organic molecules rather more effectively than it promotes their formation. It is much more likely to scorch the oceans than to fill them with life. UV is a blitz.'

However, there is a possible, relatively recently discovered energy source in the hydrothermal vents that pump out heat energy into the oceans. But just like the ability to produce amino acids, energy of itself was not enough.

Becoming complex

The problem that biologists faced in explaining the emergence of life was similar to that of cosmologists asked to explain how the universe emerged from the void. Once you have the initial step for the universe that sets up the natural laws and spacetime, everything else can be worked out in a reasonable, systematic way. In the case of biology that key hurdle seemed to be the one that provided the first life, though as we will soon see, it was not the only major step change that biologists would have to explain. The formation of life, like origins of the universe from a void, is a discontinuity that we are still at a loss to explain. This doesn't mean that it will never be possible to do so – just that we are yet to find any good reasoning behind that remarkable jump from a collection of inanimate chemicals to a structured, living cell. And then things got even stranger, because there was a second leap from the simple structures of bacteria to the far more complex cells that most other life is based on.

As far as we are aware, the additional step to complexity happened only once. All complex life is based on a sufficiently closely related set of mechanisms that it seems highly unlikely that we don't all have a single common ancestor, as noted above. Admittedly, it is possible for similar biological structures to evolve independently. The eyes of squids and octopuses, for example, are remarkably similar to those of mammals, yet there is irrefutable evidence that the two types of eye evolved separately. There just happen to be only a limited number of ways that working eyes can be formed within the constraints of living creatures on the Earth. However, it's different with the fundamentals of life.

The similarities between the living cells that make up different species based on complex cells are remarkable. Nick Lane points out that it is practically impossible for anyone but an expert to distinguish between a human cell and a mushroom cell when looking at it in detail through a microscope. Not only do we share a common ancestor with a mushroom, rather than being from a separate type of first life, but it is clear from the similarities across such a wide range of species that that first common ancestor already had a complex cellular structure. The complex cell was not a trivial structure that you could imagine somehow self-assembling from a bunch of chemicals (and even that looks virtually impossible), it was an intricate collection of molecular machines, kept together by a sophisticated membrane.

In building life into our model universe, then, we have to construct, or at least to imagine, a mechanism for not one, but two remarkable steps. First, around 4 billion years ago, just half a billion years after the solar system formed, we need life to start from very simple compounds and then, after another 2 billion years have elapsed, we need a complex cell to form. For that matter, we are also unsure whether life on Earth is close to unique in the universe or a common occurrence whenever the conditions are right.

In much of science we have alternative theories available. For example, the big bang is just one of a number of theories to explain the mechanism of the earliest moments of the universe. However, it is often the case, as is true of the big bang, that one theory is the best fit to the data we have at the moment, and until new data comes along there is no reason for bringing another theory to the fore. However, the position is totally different with the origin of life. Having no mechanism to build a prediction on – and only one known example – we are left with a vast range of possibilities for the universe at large, from life being such a rare and unlikely phenomenon that ours could literally be the only example ever, through to life being such an easy thing to kick off that, provided the basic building blocks are present, we should expect life to spring up at every possible opportunity, so the rest of the universe should be teeming with life.

Where is everybody?

In reality, it now seems highly unlikely that intelligent life with high-technology civilisation is common. We can see this in two ways. One is the so-called Fermi paradox, which emerges from a passing comment by the nuclear physicist Enrico Fermi. After his participation in the Manhattan Project during the Second World War, Fermi was working at Los Alamos in New Mexico. In the canteen over lunch, he was discussing the recent outbreak of interest in UFOs (this was 1950, when UFO fever was at its height), and Fermi was supposed to have suddenly burst out: 'Where is everybody?'

His point was that, given the apparent likelihood of life emerging (the statistics were significantly more pinned to the optimistic end of the spectrum back then), you would expect our local arm of the Milky Way galaxy to be teeming with life, and we should see plenty of visitors, with clear, well-documented encounters rather than the dubious, vague reports that typified ufology. We now appreciate better that, even if the universe were teeming with life, almost all of it might be the equivalent of bacteria, without ever making the leap to complexity. And even if there were plenty of intelligent civilisations out there, the scale of the universe means that we could easily miss out on visitors. In reality, we are yet to discover any certain evidence of life outside of Earth, even though there are a few possibilities remaining for simpler forms in outposts of the solar system.

The other reason it seems likely that life does not have a habit of popping into existence just because the necessary components are present is that we have no evidence whatsoever of it happening more than once in the 4 billion years that there has been life on Earth. Yes, complex cells are very different from the bacterial equivalents, and we don't have a mechanism for this leap occurring, but there is no suggestion that complex life evolved entirely separately from bacteria and their cousins the archaea.

As we have seen, there is very strong genetic and mechanism-based evidence that everything living on Earth comes from the same, single source of life. If it were so easy for life to start, we have to ask, why hasn't it happened multiple times? If life could occur spontaneously in the first 500 million years, why never again in the following 4 billion? In such a scenario, we too can ask with Fermi, 'Where is everybody?' Why aren't there many different strands of new life on Earth?

As we have only the one known instance of crossing the boundary from 'not alive' to 'alive' in 4 billion years – nearly a third of the entire lifetime of the universe – it isn't too big an assumption that this is a pretty difficult step to take, making it likely that life is relatively unusual in the universe, though not, of course, giving any certainty that life on Earth is unique. The truth may be out there, but if it is, it is likely to be very thinly spread.

The mystery of the past

Just as with our attempts to look back to the beginnings of the universe, we are hampered in looking back to the origins of life on Earth by the inability to access direct evidence. Biologists have it even worse, in fact. The astronomers and cosmologists researching the early origins of the universe have one huge advantage over their palaeobiologist colleagues attempting to fathom how and when life came into being. The astronomers have access to a handy time machine, powered by the speed of light, that enables them to view the past.

Whenever we look out into the universe we are looking back in time. Light travels at around 186,000 miles per second (300,000 kilometres per second), so we see the Moon, for instance, around 350,000 kilometres away on average, as it was about a second ago. Light takes eight minutes to reach us from the Sun and around four years from the next nearest star other than our own. By the time we take a look at the Andromeda galaxy, which is just about the most distant object visible to the naked eye, we are seeing 2.5 million years back in time.

With modern telescopes, astronomers can take that view back around 13 billion years. But when we look at the Earth we are limited to seeing it as it is pretty much now. Even if there are direct relicts from the early years of the Earth, they will have undergone vast processes of change in the intervening four-plus billion years. The view into the past is extremely indirect, leaving ample and inevitable room for misinterpretation.

When I was at university, I bought a lovely book called *Motel of the Mysteries*. In its graphic novel-style adventure, the author, David Macaulay, features an archaeologist, Howard Carson. Just as Howard Carter uncovered Tutankhamun's tomb, Howard

Carson, who is from the year 4022, uncovers a structure long buried in the American desert: the Toot'n'C'mon Motel. Inside a preserved motel room, Carson has to interpret finds given his own limited knowledge of twentieth-century culture. So, for instance, a TV set becomes an altar, while the toilet seat and the 'sanitised for your protection' seal on the toilet become a sacred collar and headband. Macaulay uses humour to make a serious point about the ease with which we can misinterpret ancient finds without appropriate context.

At a talk I gave on time machines recently, someone in the audience asked if we might somehow see the light from a past Earth reflected back from something shiny at a great distance and so be able to use the time-delay effect of light to have a window into the Earth's past. It's a beautiful idea, but a time mirror would suffer from severe difficulties. Getting a clear image of anything as dim as the Earth many light years distant is almost impossible, because of the relatively few photons that will get through to that distant location and all the possible distortions caused by intervening material. So, lacking this imaginative tool to see into the past, the palaeobiologist has to make use of indirect mechanisms that make the *Motel of the Mysteries* deductions seem straightforward.

Even palaeontologists making pronouncements about dinosaurs, looking back a mere 100 or 200 million years, have to incorporate a considerable degree of speculation into their work. We are dependent on the fossil record, which is both hugely incomplete – it is in reality very unlikely that an animal will end up preserved as a fossil – and involves a process that preserves only limited aspects of the animal and plant. A good example of the degree to which educated guesswork has been involved is the relatively recent transformation in our ideas of what creatures like the velociraptors featured in the *Jurassic Park* movies looked like. The lizard-like skin adopted by Hollywood had been the standard assumption for decades. But according to modern palaeontologists, these dinosaurs were almost certainly covered in feathers. Not surprisingly, the recent *Jurassic Park* films have ignored this development. The colourful feathered velociraptors might still have been vicious killers, but would be far too cuddly to look it.

A crystal ball

However, those looking for evidence of the beginnings of life envy the ease with which dinosaur hunters make their pronouncements. The ancient life experts have a much harder task because there is no direct equivalent of a fossil for a bacterium. There is no solid structure to survive; nothing more than chemical deposits can remain from those earliest organisms. So it is remarkable that we can even have a good guess that life started around 4 billion years ago. And the clue comes from small crystals of zirconium silicate and deposits of carbon that form part of ancient sedimentary rocks.

The crystalline zirconium silicate structures (known as zircons), some of which have been dated as over 4 billion years old thanks to uranium decay dating, have a tendency to trap other particles present when they form, and give us our best evidence of what the environment was like on Earth when life is thought to have begun. Unlike the organic soups proposed in the 1950s, the actual chemical make-up of the atmosphere of the period seems to have been more like the present mix (apart from a notable lack of oxygen, which was a contaminant produced by early life forms): an atmosphere dominated by nitrogen, water vapour and carbon dioxide.

The evidence for the beginning of life itself comes from carbon deposits within the zircons. Carbon comes in two stable isotopes – variants with different numbers of neutrons in the atomic nucleus – carbon 12, the more common version, and carbon 13. There is also the radioactive isotope carbon 14, used in radiocarbon dating, but in the zircons what is observed is that the mechanisms used to build living organisms have a slight preference for the smaller carbon 12 atoms, as a result of which, life tends to accumulate collections of carbon 12 that differ from the usual proportion of carbon 12 to carbon 13 of around 99 to 1.

And that's as good as it gets. We think life was active by 400 million years into the existence of the Earth because there is more carbon 12 in these zircons relative to carbon 13 than we would normally expect. There is no doubt that there could be other reasons for this accumulation, but the hope is that it was the residue from the earliest life-forms, just as the carbon deposits of coal are the remains of fossilised plants. The most recent findings in 2015 put the best estimates for that first life on Earth at around 4.1 billion years ago, well before the bombardment of the inner parts of the solar system that left the Moon in its dramatic cratered condition.

Technically speaking it's the zircons that are 4.1 billion years old, while the carbon is older still, as it was already there when the crystals formed, but we don't know how much older the carbon is. However, if this indirect evidence is correct, it does leave us with life beginning less than 400 million years after the Earth formed – a remarkably early start, especially given the way that bacterial life has stayed pretty much structurally the same ever since, not changing dramatically in form for 4 billion years.

It's not that bacteria don't evolve. With their quick reproductive cycle and ability to swap genes between individuals, bacteria are fluid organisms indeed. This is how they manage to become resistant to antibiotics. However, there have been no real changes to their structure and form. A bacterium back then would be recognisably the same thing as a bacterium now. When you consider that it has taken only 75 million years for human beings to evolve from our common ancestor with mice, the ability (or limitation) of the bacteria to stay pretty much the same for 4 billion years is more than striking. It implies that something very special happened when the complex cells that all multi-cellular life is based on were formed.

Living structures

By around 3.2 billion years ago, structures were beginning to form that seem closer to fossils, particularly in large clumpy formations that resemble the colonies of archaea or bacteria called stroma-tolites that are still found living today. There is also increasingly convincing evidence in mineral deposits from this period, strongly suggesting the impact of living organisms and of photosynthesis, with its potentially destructive production of the oxygen that would enable the development of more modern forms of life.

The assumption has always been that getting life started in the first place was the biggest challenge to explain, but in some ways that leap 2 billion years later to complex life was just as remarkable, which is one reason why an increasing number of observers suspect that, though there may be plenty of examples of simple life in the universe, complex life is likely to be far less common. As most of us first discovered in biology at school, the complex cells found in all multi-cellular life forms are the result of a kind of symbiosis, where two initially separate organisms interact for their mutual benefit.

The mitochondria that are the 'power cells' of complex cells, and the additional chloroplasts handling photosynthesis in plants, are almost certainly derived from bacteria that were somehow integrated into another single-celled organism. Developing this way still requires a frame of reference for the symbiotic pair to survive and thrive under natural selection, though it's not one of the common mechanisms that drive evolution. But to think that this coming together explains how the whole complex cell is formed is a bit like assuming that if you understand the electrical battery in a petrol car you know how the whole car works.

It is now almost certain that the predecessor of the mitochondrion – that cellular power supply – was first absorbed into an archaeon. Archaea are single-celled organisms that appear at first sight to be similar to bacteria, and that were first thought to *be* bacteria, but are now recognised as entirely separate types of organism. Archaea have totally different internal mechanisms to bacteria. Specifically, the workings of archaea have more in common with the equivalent mechanisms in the complex cells known as eukaryotic cells that form all complex life. This single starting point, however, is a long way from explaining how the merger of two very simple forms of life resulted in the evolution of the fantastically complex modern eukaryotic cell.

For many years, biologists assumed that a whole range of different components were added to a simple cell like a bacterium, building up the structure piece by piece to eventually form the complex assemblies within the cell that we now see shared by animals, plants, fungi and algae. This could have happened, it was argued, by the usual processes of evolution by natural selection gradually adding a piece here and a mechanism there. Or, in a more radical suggestion by Lynn Margulis, the originator of the mitochondrion-as-symbiote theory, it could have been as a result of extra symbiotic relationships forming, gradually adding abilities by the merging of the growingly complex cell with a simpler organism that already had the required machinery. But neither of these turned out to be a good enough fit to the current form of eukaryotic cells.

If either of these pathways for adding mechanisms had been followed, we might expect that the various forms of complex life had a number of starting points, just as complex features that have evolved like eyes do not have a single origin. If it were possible to add the kind of complexity found in eukaryotic cells by gradual evolution or symbiosis, it would be surprising to find that a single path had been taken – that smacks of the development being directed, rather than the blind progress of the evolutionary process. However, it became increasingly clear as scientists were able to examine and compare the genetic and molecular make-up of different organisms that there were not multiple versions of complex cell machinery.

After considerable research it was shown that every one of these hugely varying organisms that have been studied, ranging from humans to mushrooms to algae, originated in a single common ancestor. An organism that *already* had in place the vast majority of the complex mechanisms and structures we find in the cells of this dramatically diverse population. As Nick Lane puts it: 'The killer fact that emerges from this enormous diversity is how damned similar eukaryotic cells are.' All that way back in time, stretching 2 billion years or so, our common ancestor was already extraordinarily complex in structure.

Of itself it wouldn't be surprising that we don't have any clear route to get from the much simpler and earlier bacteria and archaea to modern complex cells. One of the inevitable results of evolution by natural selection is that a lot of the intermediate steps are rapidly pushed out of existence by their more effective descendants. The process of selection implies selecting out the less able versions, which then disappear and mostly leave no trace. Even for more recent organisms like dinosaurs, the fossil record is somewhere between patchy and hilariously inadequate, while for single cells it is often non-existent. The very idea of 'missing links' is a joke in the field, because pretty well everything is missing – we just see the occasional snapshots that survive by happenstance.

There are a number of fossilisation mechanisms, but just think for a moment what is required to produce those familiar dinosaur fossils. The animal has to be completely covered over with some form of sediment before it can decay or be eaten. The conditions have to be right for the surrounding material to harden as the bones are gradually dissolved, forming a natural mould in which minerals can crystallise in the form of the original bones. During this process, the bones need to remain undisturbed by animals and weather. And then, millions of years later, someone has to have the luck to come across them. It's hardly astonishing that our record is incomplete.

However, what truly *is* surprising is the one-off nature of the origination of the complex cell. If it could happen once, why hasn't it happened over and over again? As we have seen, bacteria, and for that matter their distant cousins the archaea, are magnificent in their ability to keep essentially the same form for so long. All the hectic evolution they have undergone, spurred on by their extremely rapid life-cycle and ability to swap genetic information, leaves them resolutely as bacteria. They may end up as bacteria with very different metabolic processes as a result of those genetic changes – but they are bacteria nonetheless.

Bacteria and archaea have kept this way for however long

they've been around – up to 4 billion years. This is not a reason to look down on them and consider them obsolete. Rather, they have shown that they are so superbly good at what they do that they have been able to carry on far longer than any of their competitors. On a simple occupancy count, this is a planet that continues to be dominated by bacteria and archaea with relatively few other niche organisms such as humans. Bear in mind that the typical human body contains between one and ten times as many bacterial cells as it does human cells. And that's on a cell-by-cell basis. Comparing organism with organism they are ridiculously more successful in terms of their ability to populate the Earth than any animal or plant.

Bacteria and archaea are indeed brilliant at what they do – which is being themselves. But what they seem pretty well incapable of doing is becoming something else, unlike complex cells, which, as we have seen, have emerged only once in 4 billion years. And yet, despite their relative scarcity, complex cells have subsequently achieved far more than bacteria ever did. In around 2 billion years, that single type of parent complex cell has undergone a process of transformation that has brought about the existence of every alga, fungus, plant and animal. Thanks to the relativistic power of evolution, we eukaryotes are masters of change.

Life in space

Even with traditional means of evolution by natural selection and symbiosis available as mechanisms, there was still the problem for biologists of finding a route to have complex eukaryotic cells arise once and once only. One radical response to this challenge of the emergence of complexity, when bacteria and archaea are so determined to stick with a winning approach, would be to sidestep evolution and pick up on the concept popularised by physicists Fred Hoyle and Chandra Wickramasinghe, known as panspermia. Hoyle was an iconoclastic scientist who delighted in overthrowing the status quo. He originated and championed many truly original ideas – some of which have stood the test of time, and some of which have proved dead ends. Arguably this is the mark of creativity – if you never suggest anything outrageous, you aren't pushing the boundaries sufficiently to make a significant change. But none of Hoyle's passions was quite as eyebrow-raising as panspermia.

This is the idea that life did not start on the Earth, but came to us from outer space. Hoyle and Wickramasinghe did not originate panspermia, which had a number of prominent supporters in the nineteenth century, including the physicist William Thomson, better known as Lord Kelvin, but Hoyle and Wickramasinghe were panspermia's most vocal supporters and did the most to put it on a modern scientific footing. They pointed out the presence of organic material in space, detected by the optical signature of its component atoms through spectroscopy.

The pair were the first to propose that the interstellar dust found in many parts of space was mostly organic, which has since proved to be true. And unlike the non-existent organic soup needed for the early models of how life might have been generated on Earth (see page 187), we know that there are complex organic molecules, including some of the amino acids that are vital building blocks for most living organisms, already in space. And we know that various objects that have entered the Earth's atmosphere and have survived to land on the surface still have organic material as part of their make-up.

The concept of panspermia was used by Hoyle and

Wickramasinghe to explain not only the origin of life in the first place, but the emergence of some of the new diseases that seem to come from nowhere. And this 'life-from-space' theory can, of course, also be employed to explain why the development of complex life is a one-off – that singular occurrence would be accounted for if complex life were highly unlikely to have formed on the Earth and instead came to us in a deposit from the stars.

This all sounds very helpful as an explanation, but unfortunately there is little scientific support for panspermia outside of Hoyle and Wickramasinghe's followers. One obvious objection to the theory is that all it does is to push back the problem of how life, then complex life, could emerge on Earth to a space-born environment or another world (although there was, of course, far more space and time available in which a very unlikely event could occur, once we take in more of the universe). Of itself, this isn't an argument against panspermia, but it makes it clear that panspermia still doesn't explain how life started or became complex.

More fundamentally, most scientific observers would argue that there is no need for the added complexity of an explanation based on panspermia. There is, after all, no good evidence of living organisms arriving on the Earth from space, even though we have been able to examine a whole lot of incoming material in the form of nearby dust still in space and meteorites that make it to the planet's surface. It's one thing to find relatively simple organic matter, just as it is to make it artificially in the lab, but it is another to discover life on a meteorite. This is one of the reasons there is so much interest in the possibility of life on Mars, as we regularly find meteorites on the Earth's surface of Martian origin.

Easier by design?

The arguments I have been putting forward about the difficulty of complex cells emerging may seem dangerously close to those applied by followers of intelligent design (ID). ID supporters argue that there are biological structures so complex, and that rely on such a level of interaction between different mechanisms, that the ID crowd could not imagine them evolving, because any intermediate structure would not provide the desired result and so would be unlikely to survive in an evolutionary situation.

What's the use, they might argue, for half an eye or half a wing? Surely evolution would dispose of such partial forms as an unnecessary burden before they got all the way to form useful organs? ID supporters argue, therefore, that there must be some form of intelligent designer, an absolute force that can override evolution's frames of reference – for want of a better term, a god – though they are careful to avoid this wording in order for their ideas to be considered a scientific alternative to evolutionary theory. I am not supporting this viewpoint.

There are several problems with the ID position. First and foremost, it is a fundamentally flawed argument because it relies on a false either/or position. The devotees of ID tell us that *either* a mechanism can be explained by evolution *or*, if it can't, then that mechanism has to have been created by an external intelligence. However, even if evolution were not an applicable mechanism, it doesn't mean there isn't another perfectly natural mechanism that does not require us to bring in an external intelligent designer. There could be many alternatives to evolution by natural selection in a particular example, and disproving one solution does not immediately make a specific other solution true. A simple case in point is the incorporation of bacteria into simpler cells to become the mitochondrial power source. This is not an evolutionary process in the usual manner. Wherever evolution probably isn't the answer, there could be many other mechanisms. It certainly isn't an 'either evolution/or designer' situation.

I should stress, by the way, should you have a religious faith, that this argument also does not exclude the possibility of an intelligent designer, it merely states that the absence of a clear conventional evolutionary process is not in any sense a proof of design. And it seems reasonable that an extraordinary solution like ID requires extraordinary evidence to prove it. What's more, this dependence on an either/or argument is only the start of the problems for ID. Many of the examples that have been brought up to 'prove' the failure of evolution simply don't work in practice.

A frequently cited example that dates back to the early days of evolution is the development of the eye. 'What's the use of a part-formed eye?' those who doubt the validity of evolution ask. During the period that it is not fully formed, the useless not-quite-eye is an evolutionary burden, not an advantage, and is liable to be bred out of existence. This might be true if the partial eye didn't do anything. But we have examples in nature of a whole host of different versions of eyes, some of which are little more than light-sensitive patches, others using a pinhole rather than lenses, and yet more having different approaches to sophistication – from the single lens of mammals, fish and octopuses to the compound eyes of insects. A partially evolved eye can, in fact, deliver plenty of benefit to its owner.

Ah, the response comes, but what about a wing? There's not much use to a wing that hasn't developed far enough to enable flight. At first glance, this is a powerful argument. A wing is a forelimb that is no longer available for manipulation or locomotion, so that it can become an aerofoil. Something in between would be neither a good arm/leg nor a useful wing for flying. However, again, the picture is not as clear-cut as it seems. Some wings are less sophisticated than others, perhaps only capable of gliding, rather than of true 'powered' flight – yet they still provide an evolutionary benefit.

More importantly in this case, it is also possible that the benefits of an intermediary stage of development could be different from those that eventually emerge with a functional wing. It's important to remember that a central tenet of evolution is that it is not directed. It has no goal or idea about the future. Nothing is trying to achieve a way to fly when wings evolve. If, for instance, a mutated limb that is part-way to a wing might give an animal some benefit in self-defence, or in the ability to store nutrients, or to radiate off excess heat – or a host of other possibilities – then that proto-wing might thrive for that reason before it eventually evolves further to happen to become a practical wing.

Perhaps the strongest example given to support ID is the flagellum. This is a rotating, tail-like structure used by some single-celled organisms as a biological propeller to push them through water. A flagellum has a sophisticated mechanism with a kind of rotary motor to turn it, which it is hard to imagine would provide any benefit if it weren't in its final sophisticated form. But in reality, the components that make up the biological motor are not unique and appear to have already been used in other biological systems, for example that used by some bacteria to secrete attack proteins. There is no doubt that it is harder to see how a flagellum evolved than many biological structures, but that doesn't make it impossible. There is no smoking gun.

When it comes to the development of life and then of complex cells, we are looking at a bigger challenge than explaining the origin of a single organ or structure, and it can be hard to see how conventional evolutionary processes could make such a leap possible. Yet, once more, this does not inherently and automatically push responsibility into the hands of a creator. It simply makes life much less likely to occur. The development of life, and then of complex life, increasingly appears to be a much rarer event in the universe than was once thought.

Biochemist Nick Lane argues convincingly that the complex eukaryotic cell came into being as a result of an interaction between two simpler prokaryotic organisms, bacteria and archaea, with a bacterium taking on an internally hosted symbiotic relationship with an archaeon. These bacteria then evolved, losing genes to the host, into the inseparable mitochondrion. This was a rare event, and what's more the success of the resultant eukaryotic cell was dependent, Lane argues, on the necessity to have a whole mix of genetic information from different bacterial and archaeal sources, fed into the parent single cells by an unusually high level of gene transfer. It may be that only a small number of initial states would result in a successful eukaryotic cell, meaning that the emergence of complex life is like winning a lottery that has a vast number of possible tickets – incredibly unlikely, but still likely to happen somewhere as long as there are sufficient 'tickets' in the form of appropriate environments and conditions.

That lottery ticket combo of bacterium and archaeon was missing the vast majority of the structures we find in complex life. These remaining mechanisms would have to develop gradually by the usual processes of evolution, spurred on by interaction between the two components of the new cell, and this itself is extremely unlikely – but without that initial winning ticket there would not have been a suitable platform for this to happen. Neither a bacterium nor an archaeon alone could support this kind of development.

The special case scenario

Some scientists don't like the idea that life could be rare, for what are arguably irrational and unscientific reasons. We have already met the astrophysicist Fred Hoyle. One of Hoyle's greatest ideas, developed with colleagues Hermann Bondi and Thomas Gold, was the steady state theory of cosmology, for some time a well-supported alternative to the big bang theory. Although the theory was originally inspired by a movie which began and ended with the same scene (the excellent classic horror film *Dead of Night*), one of the driving forces behind developing steady state was that cosmologists (and Hoyle in particular) did not like the idea of a point in time for the creation of the universe, as this seemed to make a creator more likely.

This is not a logical or scientific argument, but a relatively rare example of scientists allowing prejudice to slip into their work. As it happened, the steady state theory was at odds with new data that arose in the following decades, and though it could have been modified to deal with these problems, just as the big bang theory has been updated to cope with different conflicting data, there was more support for big bang. This meant that even an enhanced quasi-steady state that could still be an effective theory was sidelined. So it was the theory that fewer cosmologists found uncomfortable that won the day.

Similarly, a number of scientists have a problem with the idea that life is a rare occurrence, because this makes the Earth a special place, and it is statistically unlikely that there should be anything special about our planet. The doubters think that any special privilege awarded to the Earth breaks the 'Copernican principle', named for Nicolaus Copernicus, though the principle was first stated not by the Polish astronomer but by Hoyle's collaborator Hermann Bondi. The principle is so called because Copernicus was one of the first to argue that the Earth was not the centre of the universe. The principle is only a rule of thumb – it clearly doesn't *have* to be true – but it seems reasonable.

When considering the scarcity (or otherwise) of life, however, the Copernican principle is on shaky ground logically. After all, logic says that we can only make this observation from a place where life exists, and if it's rare then so be it. But the low probability irks those who, perhaps, aren't totally comfortable with probability and statistics and for them it raises a kind of prejudicial dislike. If we lay that aside, the lack of multiple sources for complex life on Earth does indeed suggest that intelligent life is, at the very least, a rare occurrence in the universe.

Despite this, our existence also suggests, interestingly, that there may be plenty of other planets where life, or even complex life, *could* have emerged but didn't. With a single lottery ticket-holder, the chances of a win ever happening are negligible. In the UK, there is now only a 1 in 45 million chance of winning the main National Lottery draw. Buy a single ticket for each draw and, on average, you will wait around 212,000 years to win. Yet in reality there are still regular jackpot winners throughout a single year, because there are lots of tickets bought each week. It's much more likely there will be a win if there are plenty of ticket-holders, and similarly it's much more likely that there will be life *somewhere* if there are plenty of planets with the right conditions.

However it happened, both life and complex life did emerge here, and once it was available, the evolutionary response of life to its frame of reference provided the mechanism to move from a world with absolutist tendencies to one where relativity holds sway. We've seen how matter and time introduce relativity to space: life introduced relativity to atoms and molecules.

Take a look at those requirements for life (page 185) again. The only one of them that could work in principle without a frame of reference is reproduction. Organisms that can clone themselves reproduce without interaction with other members of the species. But to have a mechanism for variation and to have competition requires interaction. Competition is inherently about the relative abilities of two members of the species, influencing the probability of carrying forward those abilities (or more accurately the genetic make-up that provides those abilities). The 'natural selection' part of evolution by natural selection inherently requires a frame of reference.

Relativity is also responsible for doing away with one of the apparently stronger creationist arguments against evolution – the species paradox. Some creationists, and certainly the intelligent design branch, would argue that they accept the basic principle of evolution. It's practically impossible to argue against it once you accept that those mechanisms are present. But, they would say, evolution happens only at a micro level. It can explain how, for instance, we can start with a basic dog and end up with the vast panoply of variety that we find in dogs, because we can selectively breed for specific characteristics. But a dog remains a dog.

No matter how much visual difference there is between the Great Dane and the Chihuahua, they are both dogs. They could in principle, if not necessarily in practice, breed. They remain genetically dogs. 'We're happy with that', say the creationists, 'but despite all that variation, there is no new species here. Evolution doesn't explain where species come from. It doesn't allow us to cross the species divide.' (Getting from wolves to dogs is perhaps more difficult for them, but they might argue that the division is arbitrary, because wolves can breed with dogs.)

The problem of understanding largely arises from the arbitrary nature of the label 'species'. Admittedly it's difficult to see at first how new species could ever evolve, because, despite evolution, every organism produced by a normal breeding mechanism is the same species as its parent. That would seem to imply that it is indeed never possible to produce a new species by evolutionary processes. The paradox is that, given enough generations, you can do just that. Even though every organism is the same species as its parents, it can be a different species to its ancestors or descendants.

A useful parallel is to think of a rainbow. Forget the paltry seven colours that Newton dreamed up and that we still parrot today – let's divide that rainbow up into, say, 16.7 million slices. That's similar to the number of colours my computer screen routinely handles. If I go into an image program, I can select the red, green and blue components that make up the colour of light in each pixel, with 256 variations of each. That's a total of 16,777,216 ($256 \times 256 \times 256$) available colours.

Now let's take two adjacent slices out of my 16.7 million. Would you describe those two slices as the same colour? Absolutely. Of course, we know that there is a tiny variation, but it would be impossible to name that as producing a different colour. And that's true of *any* pair of adjacent slices selected throughout the rainbow. So I have a series of slices, each the same colour as its neighbour, and yet as I move through the series along the whole rainbow, I will pass from red through orange and yellow and green, all the way to blue, indigo and violet.

The same thing occurs with species. Each organism is the same species as its parent as you head back through its ancestry. Yet if you go far enough, you will find that you are looking at a different species to the one you started with, all the way back to that common ancestor for animals, plants, fungi and algae we met earlier. We can do this because of relativity. The concept of species is not an absolute one, but a relative description. The frame of reference of the term 'species' can only ever emerge from a specific generation. Given a particular example of an organism we can extend a bracket saying 'same species' back a number of generations – but there is no absolute concept of species that defines a fixed set of generations. Evolution produces not just variations within a species but also, over time, new species – because a species depends entirely on the frame of reference of an individual.

◇◇◇◇◇

When life does occur, then, we can be sure that relativity is at play, and not just in assessing our existence with respect to a relatively unpopulated environment. At a basic level, that relativistic evolutionary interaction applies to all living things. And a combination of life and evolution can deliver remarkable developments, none more so than the capabilities of the brain and of consciousness demonstrated so dramatically by the human species. These are biological developments. They all belong in the chapter on life. Yet the human mind also provides a whole new opportunity. We have found it possible to go beyond nature's capabilities and bring creativity and innovation into play, which enables our species to override life's frameworks.

8 Creativity and Innovation

◇◇◇

There is one final layer to add to our DIY universe if we are truly to understand humanity's place in it. With space, stuff, time, motion and gravity we can incorporate most simple physical structures. And life brought in many of the things that we can identify on Earth as being more than just matter and light. However, echoing *The Ascent of Man*, there is something further required to set the reference frame for humanity itself.

Exactly what this is has always proved difficult to pin down. Across the years it has been suggested that the thing that makes humans special is writing, or speech, or art or play. It clearly isn't just having a mind – that's a label we can apply to many living things. And it is very difficult to be sure that 'intelligence' will suffice. There are plenty of examples of animals showing forms of intelligence. I first tried to conceive of this characteristic as 'enhancement', meaning the ability to go beyond biological capabilities and biological evolution to make the human form capable of more than its biological limits. But that too proved insufficient. It became clear that the distinction I was searching for was creativity and innovation.

Changing frames

Being creative requires a special use of frames of reference. We've seen how relativity comes into play in the physical frames of reference used with everything from movement to gravitation. And we have explored the way that life depends on the environmental frame of reference that drives evolution by natural selection, where one organism has a better fit than its competition. But by undertaking a creative act we do something unique. Consciously or unconsciously, we *change* frames of reference.

Edward de Bono, who devised the concept of lateral thinking, describes how those seeking to make something happen are hampered by tunnel vision. Imagine we need to solve a problem or come up with a new idea. In principle we have access to every possible idea. Imagine a three-dimensional space – 'idea space' – filled with vast numbers of stars, each of which is a possible idea. We have a huge range of possibilities open to us, but we rarely make use of that expanse of creativity.

In our day-to-day lives, we live in a tunnel within this space, forced on us by our assumptions. These are assumptions about our environment, what is possible, how things have always been done and so forth. This tunnel limits our capabilities to be creative and to innovate effectively, reducing our access to a small portion of the overall idea space. De Bono suggests that we should use 'provocations' – these are techniques (many of them originally devised by advertising guru Alex Osborn) that are designed to thrust us out of the tunnel to a different starting point in the idea space. From this new viewpoint, we can then produce genuinely original ideas.

What de Bono is describing is a method to force a move to a new frame of reference. We traditionally view the world using a particular personal frame, based on our assumptions. The provocation technique pushes us to look from a new direction, with a new reference frame. And it is by seeing the world this way that we can come up with original thinking.

Of course, people had ideas long before practitioners like de Bono and Osborn came along. What the creativity experts added was a collection of techniques to make shifting our frame of reference easier to do. Being creative and innovating is something that humans have employed throughout history. And arguably we are pretty much unique in this. To quote psychologist Mihaly Csikszentmihalyi: 'Without creativity, it would be difficult indeed to distinguish humans from apes.'

Changing the reference frame is central to any creative act. Physicist David Bohm, writing on creativity, suggests that the ability to change our mental frame of reference is a continuation of a child's ability, which some people maintain into adulthood far better than others:

> There must have been a considerable body of scientists who were better at mathematics and knew more physics than Einstein did. The difference was that Einstein had a certain quality of *originality*. ... [A] child learns to walk, to talk, and to know his way around the world just by *trying something out and seeing what happens*, then modifying what he does (or thinks) in accordance with what has actually happened. In this way, he spends the first few years in a wonderfully creative way. ... As the child grows older, however, learning takes on a narrower meaning. ... So his ability to see something new and original gradually dies away. And without it there is evidently no ground from which anything can grow.

It might seem that while the reference frames of, say, special relativity are a constituent of reality, the ability Bohm describes is imaginary. It is 'just' the way that we see the world. But whether the frame is the way that physicists choose to consider light – a wave, a particle, a disturbance in a quantum field – or the way that artists see what's around them, or that businesspeople come up with a new product, it is still all about viewpoint. Each involves changing a frame of reference and hence is inherently relativistic.

Bohm again: 'In [a creative act] I suggest that there is a perception of a *new basic order* that is potentially significant in a broad and rich field. This new order eventually leads to the creation of new structures ...' The innovation that resulted from this kind of creative thinking has been responsible for transforming humanity, over a biologically brief period of tens of thousands of years, from just another ape to something radically different.

The evolutionary race

In purely biological terms, *Homo sapiens* has changed very little in the last 100,000 years. But to take such a limited view misses the fundamental importance of the impact of our creativity, something that has proved surprisingly easy for some observers to do. This was clear from an article in a scientific journal claiming that chimpanzees are 'more evolved' than human beings. To be fair, the author qualified this statement by putting the word 'evolved' into what I'd describe as embarrassment inverted commas, and by saying that it's only in one sense that this is true, but the point was still made.

The article was the outcome of work by scientists at the University of Michigan in Ann Arbor, comparing 14,000 genes that are shared between the human and chimpanzee genomes. Chimpanzees are our closest living relatives, and what was fascinating was that of the genes compared, 233 of the chimp's had changed in a way that suggested natural selection had kept the change because it gave benefit to the species, while this was the case with just 154 of the equivalent human genes. The lead researcher from Michigan, Jianzhi Zhang, commented: 'The result overturns the view that, to promote humans to our current position as the dominant animal on the planet, we must have encountered considerable positive selection.'

It's important not to read too much into gene comparisons. The difference between organisms with shared genes is often down to combinations of genes and external epigenetic processes where biological and environmental factors result in different genes being switched on and off. Even where the individual genes may be significant, it's not practically possible to pin down, for instance, the genes that mean we have bigger and more powerful brains than a chimpanzee.

What's more, the study was only able to compare a very small section of the comparable genomes of the two species. Despite this, the article concluded with a remark from Victoria Horner from the Yerkes National Primate Research Center in Atlanta, Georgia, that showed the shortcomings of taking a purely biological viewpoint that is prepared to ignore – or represents a lack of realisation of – the importance of the ability to be creative: 'We assume that chimpanzees have changed less than us when that's actually not the case.'

The problem with basing the degree of change a species has undergone on a purely biological, genetics-based viewpoint perhaps illustrates the over-emphasis of the importance of genes in public science communication, a position that has emerged since the publication of books like Richard Dawkins' *The Selfish Gene*. First of all, as Jianzhi Zhang stated in the article: 'It is possible that the genetic changes underlying brain size are very few.' Simply counting genes that have been modified is not an effective measure of the way that an organism has developed. Sheer numbers of genes don't in themselves give a useful picture of the complexity of an animal or plant.

Like many other relatively simple organisms, for instance, the rice plant has significantly more genes than a human being, but this tells us nothing about the organism's capabilities. A small number of genes can be responsible for a phenomenally important difference in an organism, with the importance of our large brains as a prime example. Not all genes are equal in their impact.

Secondly, and probably more importantly, to say that in the last 6 million years, chimpanzees have changed more than human beings does not give a comprehensive view, but rather one limited to the specific study of genetics. In that time chimps have carried on doing what chimps have always done with very minor differences. They do not have the same ability to reframe problems and the world around them; they can't undertake scientific experiments, develop theories or innovate with technology.

Creativity is not the exclusive territory of human beings – but the significant difference is that the resulting innovation is an essential to define the human place in the universe while it has little impact for the other animals. Take the strongest example of innovation outside the human sphere: tool use. A number of species from those chimpanzees to some birds, notably the crow family, have been observed using tools to extend their basic biological capabilities. These include the use of stones to crack open shells and 'fishing' for insects by poking lengths of grass or twigs into logs. Such tool use is limited, not always universal within the species that uses it, and has not made a significant transformation of the way that these animals live their lives.

In such circumstances, there is considerable doubt that the innovation that has taken place is a result of a species reframing its view of the world. Such examples become trivial when set alongside the creativity that has taken humans from being just another ape to the animals that have transformed the Earth and added something that life, left to its own devices, never could. Unlike every other organism, we have the potential to choose to spread beyond our planet of origin – and without creativity, this could never be more than a dream.

Making the shift

To quote physicist David Bohm again:

> In [the creative frame of mind] one *does* something (perhaps only to move the body or handle an object), and then one notes the difference between what actually happens and what is inferred from previous knowledge. From this difference one is led to a new perception or a new idea that accounts for the difference. And this process can go on indefinitely without beginning or end, in any field whatsoever.

Each time, the frame of reference shifts:

> For as long as the individual cannot learn from what he does and sees, whenever such learning requires that

he go outside the framework of his basic preconceptions, then his action will ultimately be directed by
some idea that does not correspond to the fact as it is.

In other words we need to learn from our environment to be
able to shift our frame of reference, or we will incorrectly interpret what is happening.

One of the fascinating aspects of creativity that makes it feel
quite different to many of the physical processes we have met
so far is the feeling that it often involves an instantaneous shift –
that the creative individual has seen things in one particular way
for a long time and suddenly switches to a different reference
frame, making it possible to have a totally new understanding.

The astronomer Vera Rubin described this shift of viewpoint
in her discovery that some stars rotate around a galaxy in the
opposite direction to the others. She had a first hint of this from
two early spectroscopic images, but one of these wasn't very
good, and she felt that she needed confirming evidence, which
wouldn't be available for at least a year. Rubin commented:

> I sit in front of this very exotic TV screen next to
> a computer; it gives me the images of these spectra
> very carefully and I can play with them. And I don't
> know, one day I just decided that I had to understand
> what this complexity was that I was looking at and
> I made sketches on a piece of paper and suddenly I
> understood it all. I have no other way of describing
> it. It was exquisitely clear. I don't know why I hadn't
> done this two years earlier.

In addressing the importance of the shift of frame of reference, it is important to emphasise that the whole span of creativity and innovation is not encapsulated in that act. Rubin's achievement has to be seen in the context of a long period of time before and after that shift, assessing the data, making measurements and confirming results. During that time, she will have been both consciously and unconsciously mulling over her data. Many of the ways that creativity has changed the nature of humans take a long time to mature – but the change in frame of reference usually takes place quickly.

Interestingly, there is uniform acceptance that one of the best ways this switch of frame can be encouraged is to stop thinking about the topic, leaving the requirement to the unconscious mind to process. Electronics pioneer Frank Offner summed up the process:

> I will tell you one thing that I found in both science and technology: if you have a problem, don't sit down and try and solve it. Because I will never solve it if I am just sitting down and thinking about it. It will hit me maybe in the middle of the night, while I am driving my car or taking a shower, something like that.

When we consciously work on a problem or new idea, unless we use an explicit technique to encourage our brains to come at it from a new direction, we inevitably approach the requirement through the frame of reference we associate with this area. What is enabled by giving time for background processing, not explicitly thinking about what is needed, is the opportunity for the subconscious to try out different reference frames that the conscious mind would never consider.

Psychologist Mihaly Csikszentmihalyi is best known for originating the concept of 'flow', the application of a focused mental state to an activity. He suggests one possible mechanism for what is happening when we stop consciously pursuing a solution is that ideas from the thinker's knowledge domain – the stuff they already know – can combine randomly and that connections the conscious mind would dismiss are allowed to be followed up by the less rational, and hence less censored, subconscious. He also suggests there may be some parallel processing going on in the brain during these periods where mental processing isn't under conscious control, enabling a much richer space of possibilities to be considered before the thinker comes up with a new way of looking at their need, seeing it in a different frame.

Csikszentmihalyi makes the point that creativity requires knowledge of a domain – an area of expertise such as maths, or painting – a new way of looking at that domain (in my terms a change of frame of reference) and acceptance of the new idea by those in the field. He argues that it isn't sufficient to come up with a new frame of reference; it also has to be accepted by your peers. So, he argues, figures like Bach in music and Mendel in biology could not be considered truly creative at the time they worked, but became so decades after, once their work was recognised and their new way of looking at things was accepted into the domain.

There is no magic wand for creativity. Adopting a different frame of reference does not inherently produce a better view – and often it can come up with a distorted one, perhaps because the individual's knowledge of the domain is limited. As a science writer I get sent many suggestions as to why Einstein was wrong about something (he is far and away the favourite target), how to solve Fermat's last theorem in a single page, or how electromagnetism and the mystical effects of crystals are related. Each of these people has had that opportunity to see something differently, but without the appropriate domain knowledge, the result can often be pseudo-science at best.

It isn't that people who lack the appropriate knowledge can't have good ideas. An example from product development is the invention of the Polaroid camera. Edwin Land, who was a scientist and engineer, knew perfectly well that it wasn't possible to just take a photograph and see it straight away. You had to develop the film, make a print and so forth. But his young daughter, after a trip to the country, pulled the film out of Land's camera and unreeled it to look at the pictures, ruining the film. Land was then inspired to make his daughter's concept possible. She came up with the idea because she didn't know what *wasn't possible*. Domain experts often know too much about what isn't possible to come up with ideas. They need something like Land's daughter to force them to take on a different frame of reference. But then, expertise is needed to sift out the ideas that are worth carrying forward and to turn them into reality. The 99 per cent perspiration (versus 1 per cent inspiration) that Edison identified with creativity is just as important in changing humanity's place in the universe as is the original idea – arguably, more so.

Before looking at a few of the ways that human creativity has transformed our existence, and so showing why we need it in our model universe, it is worth highlighting one last aspect of creativity brought out by Csikszentmihalyi. He comments:

> What is extraordinary ... is that we talked to [highly creative] engineers and chemists, writers and musicians, businesspersons and social reformers, historians

and architects, sociologists and physicians – and they
all agree that they do what they do primarily because
it's fun.

Creativity may have changed humanity in a remarkable way, but it seems that the main driver is usually the personal reward from undertaking the process, rather than the eventual benefits for humankind.

Csikszentmihalyi suggests that we have gained an evolutionary benefit because at least some of the population derive the reward of pleasure from their innovation. Overall, as a race, he points out, we need a degree of conservatism, to avoid repeatedly heading off in new directions aimlessly. But it has proved hugely beneficial to have some individuals who are powerfully driven by the pleasure of creativity. In pure natural selection terms, these individuals are probably less likely to survive than the rest of the population, but they enable that population to move forward and improve, meaning that over time they would tend to be protected and encouraged by their peers. As such, the benefits these people accrue for the species generate a special kind of evolutionary pressure, based not on the impact on the individual, but on humanity as a whole.

Broken ground

There are so many ways in which creativity has transformed human life that it would take a whole book to explore even a sizeable subset of them. However, to get an insight into the nature of creativity and how it often involves a change of frame of reference, it is useful to explore some of the early applications of creativity. These are often developments where we have got so used to the shifted reference frame that it can be hard to see just how significant the new way of looking at the world was. And the first involves a rock.

A human fist can do a certain amount of damage, but a fist holding a rock or a club can do much more. Weapons and manual tools soon extended our natural capabilities. Even the crudest technology transformed human capability. The difference in strength between two similar pre-human species can be measured in a few percentage points. By using weapons, the strength of the enhanced human was magnified many times over.

Unworked stones increased the power of the fist (and reduced the self-damage risked with a punch), or could be thrown as simple missiles, making it possible to attack more safely from a distance. The different uses of a simple stone illustrate how easy it is to overlook the power of taking a creative view. We can't get excited about a stone, because it's only a piece of technology in a philosophical sense.

Stones lie around waiting to be used – they are little more than broken ground. There is no craft or thought required to make them – the inventiveness comes in seeing them for what they are: in picking them up and hitting with them or throwing them. And that's where an early human undertook a substantial shift in frame of reference. That's an early step beyond nature. For millions – billions – of years before, stones on the ground had no value. They were simply part of the landscape. It took that shift of viewpoint to see a stone as a vehicle to enhance the power of a punch or to cause damage from a distance. And with that shift began humanity's distancing itself from the rest of life on Earth.

An increased ability like this requires us to add in this final aspect of our model universe, following on from life itself. Including creativity brings our universe to the position we are used to in our own existence.

The prehistoric drone

In a sense, a thrown stone was the first remote weapon, but it had no autonomy, unable to do anything more than follow the trajectory on which it was first thrown. Now, both for military and civilian use, the drone is becoming ubiquitous, giving us the apparently novel ability to extend our reach with a semi-autonomous device. But the second work of ancient creativity we will add to the mix, introduced around 40,000 years ago, also provided a remote capability – and it's a technology we still use today, though in ways its original inventors could never have conceived. Here, the shift in frame was more subtle in impact, but more sophisticated.

As with all early creative developments, we don't know the detail of how this particular one arose, but it is possible to imagine what might have happened. Early hunters, perhaps warming themselves by a fire at night and butchering the day's kill, had a problem with scavengers – animals intent on taking away the meat the hunters had fought for. But one particular scavenger, a wolf, seemed different from the rest. It didn't attack the humans or try to take away the kill. Instead it lay quietly by the fire. Perhaps one of the humans rewarded it by giving it a little meat. Some time later, when the camp was attacked – by other humans or by animals – rather than running away, the wolf fought alongside them.

Then came that shift of frame. Up to this point, this particular wolf might have been tolerated as an oddity. Wolves were still considered predatory enemies and competition in the hunt. But seen from the new viewpoint, this transformed wolf had the potential to be something more useful even than a spear or slingshot. If a wolf could truly become part of the group, it provided a resource that complemented and extended human ability, making the group far more than it had been.

Within a surprisingly few generations the wolf would have become, to all intents and purposes, a dog.

In a fascinating experiment undertaken between the 1950s and the 1990s, Russian geneticist Dimitri Belyaev selectively bred Russian silver foxes for docile behaviour. Over the period of 40 years – an immensely long time for an experiment, but nothing in evolutionary terms – Belyaev's fox descendants began to resemble domesticated dogs. Their faces changed shape, becoming far less pointed than a typical fox mask. Their ears no longer stood upright, but drooped down onto their heads. Their previously erect tails became floppy. The animals' coats, which had been quite uniform in appearance, developed distinctive patterns and colorations. The fox-dogs spent more time playing, and expected more leadership, either from humans or adults of their species.

Over the course of those 40 years, Belyaev turned silver foxes into animals that approximated to dogs. The process of directed species creation does not have to take long. It is quite possible that a couple of human generations after that first tentative contact, the early hunters were no longer interacting with true wolves. The animals that frequented their camps would have changed manner and appearance. Their upright ears would have drooped. Their coats would have become more varied. And thanks to a change of frame of reference, new animals that enhanced human abilities had been brought into existence. At this point it might not have been an entirely separate species from the wolf, but the dog had been created.

Despite its short legs, a dog can run significantly faster than a human being (something I've been all too aware of when my dog was young and tried to run away). Dogs have a much more effective sense of smell than we do. Their jaws are notably more powerful than ours, with larger fangs, making them a more dangerous weapon than a human's comparatively puny teeth. If you consider the roles of hunting and protection – probably the two first reasons for making use of a dog – the dog owner has a formidable weapon with a reach that can extend far beyond that of a thrown spear, and that provides a confusing second source of danger for any attacker, who is forced to watch in two directions at once. What's more, the dog can roam, taking in places the human controller can't even see or access, providing a mobile warning alarm system.

Because of the pack loyalty that made them possible to use in the first place – one of the reasons that cats have never provided the same level of utility – dogs rapidly became more than tools, developing a close and complex relationship with their owners. Though now the majority of dogs are pets – extensions of the social family – some specially trained dogs still provide a whole range of remote capabilities, from management of sheep to assistance dogs for the disabled. Yet without that initial shift of frame of reference from wolf as predator to wolf-dog as part of the human group, none of this would have come about. The dog is a piece of Stone Age technology, one of our earliest examples of the way that human creativity can change our world. Developed 35,000 years *before* Stonehenge, it is still going strong and is used around the world.

Scratches conquer time and space

Dogs have proved valuable to humans, but they pale into insignificance alongside what writing has done for us. Hardly anything that we use in everyday life that makes us more than a biological organism would exist were it not for the written word playing a role in its development. In physical terms, writing can be as simple as a few marks in the sand or a blob on a piece of paper, but in conceptual terms, writing is a vehicle to free up communication in time and space, enabling us to destroy the shackles of here and now. Without writing, the development of science and modern forms of technology,* trade and literature, for instance, would be impossible.

Most animals and even some plants communicate at some level – but usually that communication is immediate and then lost for ever. Chemical signalling admittedly can last a little longer. When cats spray to mark their territory, their signals might last a week or two, but then the message has disappeared and will have to be resent over and over if the communication is to continue. Spatially, even this form of animal communication has strong limitations – a good thing for the cat, as the whole purpose is to mark out a local territory, but this limitation makes it impossible to reach beyond a very narrow neighbourhood.

---

* Clearly not all technology is impossible without writing. As we have seen, the most basic technologies date back to a pre-writing Stone Age. And even remarkable structures like the medieval cathedrals were primarily built without writing to assist those who did the construction, though the master masons presumably did make notes and communicate with writing. However, virtually none of the post-medieval enhancements, which have seen both human life and the Earth transformed, would have been developed without the written word.

Writing takes away the limits of space and time. I have books on my shelf containing words that were written on the other side of the world. I have words (admittedly in translation) written by Newton, Galileo and even Ancient Greek philosophers. Because I enjoy classic science fiction and murder mysteries, in my fiction section there are probably more communications from dead people than from the living, and very few of the books were written close to where I live. On my computer I can read an email typed in the middle of the night my time, that originated on the other side of the world. When you read these words it will be months or years after the moment (12:53pm GMT on Saturday, 24 October 2015) when I first typed them. The chances are that you are hundreds or even thousands of miles from my desk in Swindon, UK. It doesn't matter. Writing takes care of time and space.

To provide a final look back at the prehistoric creativity that underlies our current status, it would be hard to find a clearer example than the written word. Although (paradoxically) we have no record of how writing came into being, it is possible to deduce some likely possibilities from what has survived, and it seems that writing came about as a result of a number of sequential shifts of frame. The earliest example we have of what may be a kind of 'written' information is an artefact known as the Ishango bone. This is the calf bone of a baboon that has three collections of scratches on it, totalling 60, 48 and 60 markings respectively, and the bone appears to date back around 20,000 years.

It is possible that the marks were decorative, or random – though they don't give a visual impression of being either – but equally they could form a tally. Most of us will have used tallies at one time or another – perhaps counting off repetitive occurrences by making sequential vertical lines on a piece of paper, and on every fifth count, drawing across the previous four lines to make a set of five. Tallies were almost certainly the first of the predecessors to the written word, because they require only a relatively small shift of frame of reference. This is because tallies provide a mechanism to count without numbers.

Let's start with the simplest form of tally – using the fingers of a hand. Imagine, for instance, that an early hunter wanted to make sure he still had as many animal hides in the evening as were in his shelter that morning. In the morning, he made a tally of his hides by starting with an open hand and pushing one finger closed for each hide. With all the fingers down he pressed the thumb across them for the next hide – and that was all the hides. Then in the evening, he undertook the same process and came up with the same result. He didn't need to know that there were five hides – he didn't know, in fact. He had no concept of number. But he did know that he ended up with the same tally, so he had as many hides as before.

What the prehistoric hunter had done is what mathematicians would describe as the process of showing that two sets – in this case the set of the hides and the set of the digits on one hand – have the same cardinality, which is the measure of the size of a set. Mathematics doesn't require that we know how many items are in each set to check this. As long as we can pair off items in the sets in a one-to-one correspondence, one from each set at a time, the sets have the same cardinality.

Although this sounds complex when described in mathematical terms, it is the most basic step towards the written word. We simply remember that the tally hand, scratched on a surface to aid memory, corresponds to the hides. (This use of a hand appears to be where the traditional tally mark comes from – a hand with four fingers, crossed by the thumb.)

Over time, users of tallies would have noticed something interesting. Exactly the same tally hand worked for everything they want to check up on, whether it was hides, goats, children or plants. And with this came the next, arguably bigger, shift in frame of reference. If I was a trader in those times and someone wanted to trade me some hides, I could ask, 'How many hides?' and the other person could respond, 'A hand.' I now have a mental picture of how many hides. We would have gone from a tally, where there is a clear correspondence between two sets of physical objects (hides and digits), to counting, using a symbol, in this case the word 'hand', to represent a specific number of objects.

The final reframing would have come when early accountants realised that they could change a mark indicating that they had a hand of 'something unknown' to, say, specifically a hand of hides. By adding some extra mark – perhaps initially a picture of a hide – to the hand mark, they now had an explicit record of ownership or trade. A record that would endure, and that could be copied or taken to another location. They had, in a very crude form, developed writing.

Without additional technological revolutions, the written word would have continued to be a relatively localised interest, available only to the rich and powerful. But two final shifts – the invention of printing and means of electronic distribution, such as email and the internet – have turned writing (and its visual and audible adjuncts) into a most powerful mechanism for enabling human beings to modify their place in the universe.

Remembering creatively

The rock in the hand, the dog, and writing all predate any historical account of exactly how that reframing took place (though many of us have experienced the most recent major frame shift for writing as it moved into the electronic age). But it is worth taking a look at one more example as we begin to appreciate the power of a new frame of reference – and this is in the enhancement of human memory.

Although having a remarkable storage capacity, our memories have some significant limitations. In the short term, we find it difficult to hold more than around seven items in memory at once. (Try looking at this fifteen-digit number for a second or two – 427718960328758 – then look away, take in what's around you for a few moments, then see if you can recall the number.) And longer term, although we can often remember surprising details, our memory is very selective, based on links and associations rather than a structured set of criteria. We often can't remember what we need to recall, while having remarkable memory for trivia.

Historically, two approaches were taken to get around the limitations of the human memory. Writing is one. This versatile technology is not just about communication – written notes have helped support memory since writing has existed. The other, which dates back to Ancient Greek times, is to make use of memory techniques such as the 'mind palace', where items to be memorised are placed in different locations in an imagined building, an approach used on the TV show *Sherlock*. However, new approaches involve a reframing of the problem from 'How can I remember better with the mental resources I have?' to 'How can I directly enhance those mental resources to improve memory?'

When memories are formed in the brain, one protein, cyclic amp-response element binding (CREB for short) has a significant role to play, as it is used by the brain in the construction of synapses. Synapses are the tiny junctions between pairs of brain cells, and they also form the links between the nervous system and other parts of the body. Each neuron in your brain – you have around 100 billion of these odd-looking elongated cells – is connected to anything between 1 and 1,000 other cells. Children can have as many as 10,000 billion of these synapses, falling to something like 1,000 billion as we get older.

This dropping off, incidentally, does not reflect the old idea that brain cells die off gradually through our life and are never replaced. We now know that brain cells do regenerate, but the number of connections in the brain gradually reduces. We also know that memory is dependent on the synapses, and as Nobel Prize-winner Eric Kandel discovered in his work with giant sea slugs, the protein CREB appears to make it easier for memories to form. Kandel, who escaped the Nazis in Vienna as a boy to become a top scientist as a US citizen, has spent his whole working life exploring the nature of memory at the level of individual cells in the brain. After working for many years with giant slugs (which have unusually large, and thus easy to study, neurons), Kandel went on to use mice, raising the levels of CREB in their brains – the result was to produce mice with memories that were twice as good as those of untreated animals.

It might seem impossible to tell how good a mouse's memory is. You can't ask it questions as you would a human. At one time, mazes were used to test animal memory. The animals had to find their way through a maze to get to a titbit of food. The faster they learned the maze route, the faster the memory was assumed to have formed. But it has since been shown that mazes aren't a great way of establishing memory levels, because the skills needed to negotiate a maze are quite different from the act of remembering information. When we recall information, it is consciously retrieved. But repeatedly following a maze makes more use of procedural memory, the kind of memory that allows me to touch type. If you ask me where V is on the keyboard, I couldn't tell you. But ask me to type it, and my procedural memory delivers a press on V. It's the same with mice learning mazes.

To get round this, in Kandel's experiment the mice were placed in the middle of a brightly lit circular table with holes around the edge. Mice don't like to be exposed to bright light or to be in the centre of an open space, where they feel at high risk. They try to find a bolt-hole to get out of sight – but only one of the holes round the edge of the table was a way to escape: the rest were dummies. Initially the mice would randomly try holes, then they would begin to take a more systematic approach, but eventually memory began to kick in. Markings on the walls around the table showed which hole was the escape route. Although the position of the escape route changed from session to session, the marking was moved with the hole. This way, mice that remembered the connection of marking and hole would escape more quickly than those guessing at random or working systematically around the table.

The result of these experiments, and parallel work elsewhere, is the early stage of development of CREB-related drugs that may be used in the future to help with memory impairment, such as that caused by Alzheimer's disease. But drug companies are always looking for the biggest customer base, and there will always be more people with undamaged memories than patients who have suffered memory impairment. If the drugs can be shown to work effectively and safely, the real dream of the drug companies is to have a pill that, taken regularly, can be used to boost memory function in ordinary, healthy human beings.

Dr Frankenstein's mind manipulation

Enhancing the brain's capabilities directly is not only amenable to influence by chemical means. Magnetic fields may have benefit in memory enhancement and generally improving brain function. Initially there was justifiable suspicion about the technique known as transcranial magnetic stimulation, which involves using powerful electromagnets to influence the brain. It seemed too similar to the eighteenth-century fad of Mesmerism or animal magnetism (the word 'animal' referred to the spirit or 'animus', not to animals per se), which claimed to provide medical cures by stimulating the 'magnetic field' that was alleged to surround human beings like an aura. It was, not surprisingly, nothing more than a deception.

However, powerful magnetic coils have been used to stimulate the brain experimentally in the last few years to treat brain disorders and to help with recovery from strokes. The strong magnetic field induces electrical currents in the brain, which kick various neurons into action. Although there is inevitably something of a pot-luck result from a treatment which is difficult to focus on any detailed area, Fortunato Battaglia and his team at the City University of New York have shown that using this transcranial magnetic stimulation on mice increased the action called long-term potentiation that is used to store memories away.

The treatment also increased the levels of stem cells in a region of the brain called the dentate gyrus hippocampus. These cells continue to divide throughout our lives, and research at Johns Hopkins University School of Medicine in Baltimore has shown that there seems to be a connection between these cells and the ease with which we can store new memories. More research is still required, but if magnetic treatment can be appropriately focused in areas of the brain, it might be able to hold back the impact of memory-impairing diseases like Alzheimer's, and may help anyone improve memory formation.

Impressive though the results of external stimulation are, not everyone thinks that it is enough to stay outside the brain when trying to enhance its functions. Hands-on brain surgery is also an option to give the brain a boost. Experiments have been undertaken to enable implants to communicate directly with the hippocampus, the (roughly) seahorse-shaped segment of the brain that plays a major role in handling long-term memories.

In 2006, a team at the University of Southern California led by Theodore W. Berger removed a slice of a rat's hippocampus and replaced part of it with a chip, which was able to interact with the brain segment, emulating neurons, successfully processing the signals that are transmitted through the hippocampus. The chip had been under development for several years, following painstaking work on hippocampus cells, stimulating them millions of times and recording their responses. This was necessary as we don't understand how the hippocampus processes memories, so the hippocampus cells were treated as a black box, and the chip was made to mimic the response of the cells to different stimuli.

The USC team's follow-up experiments involved taking a step back from the chip itself, using a computer to simulate it, as the specially built chips were then very expensive to make. But the intention was always to move from working with extracted sections of brain to communicating with a living brain. This has now been undertaken with both rats and monkeys, making a chip act as an extension to the hippocampus.

In an experiment with rats, the signals produced when

memorising a task were captured onto a chip. The rats were then given drugs that interfered with their memories, making it impossible for them to undertake the task any more. When the rats' brains were fed with the information from the chip, they regained the memory of how to perform the task.

Berger's aim is to eventually provide implanted memory aids and to extend the coding to as wide a range of memory applications as possible. The hope is to use an algorithm that predicts and mimics the activity in the brain when long-term memories are formed, in principle to be able to repair damaged memory function. Eventually, Berger hopes that his work will be able to enhance human memory where brain injuries have reduced the ability to form such long-term memories.

The move from rats to humans would lead to problems, over and above the obvious issues of risk to the subjects. The implant has to be able to model the action of the neurons in the brain, but there is currently no way to do this without disrupting brain function. We have no way to safely scan brain signals at the level of individual neurons without being intrusive, which has meant that researchers have suggested they may have to use models of monkey neurons for human trials. However, it is possible that non-intrusive scanning at this level of detail will be available at some point in the future.

Memory enhancement chips like this were designed to repair brain damage, enabling memory function to be restored when it is failing, but they could also conceivably boost the capabilities of a healthy brain. Although the brain is an incredible, vastly complex structure, neural mechanisms are very slow compared with electronics – a memory-boosting chip could, in principle, make memory much faster and more efficient for the kinds of simple factual storage that human memory struggles to achieve.

When IBM's Watson computer won the US TV *Jeopardy!* quiz game in 2011 it demonstrated the superiority of silicon for information recall. In principle a sufficiently advanced chip could give the human brain the same ability.

Some futurologists love the idea of the wired human, the person with the socket in their skull to jack into the electronic world and expand their mind. This was seen in dramatic form in the *Matrix* movies and variants have cropped up in countless science fiction novels. However appealing the benefits, though, it seems hard to believe, with our natural and very reasonable squeamishness about the brain, that many of us would allow ourselves to be tampered with at this level just for fun. One thing is certain – if wired connections to the brain ever did become a commonplace reality, those frightening-looking connections in the head we see in the movies would be non-starters, especially the huge, unsubtle sockets that featured in *The Matrix*.

The biggest problem with introducing electrodes into the brain, apart from the non-trivial risk of damage during implantation, is that the point at which the wire (or socket) passes through the scalp and through the skull is a potential source of infection that would constantly put the brain at risk. Should wired brain interfaces become commonplace, they would be located wholly under the skin, using non-contact methods of communication, like the RFID (Radio Frequency IDentifier) tags frequently used now in stock control and to make payments by mobile phone. However much we move away from the unsightly and dangerous sockets in the head, though, greater hope must be held out for developments in external, non-intrusive electronic brain interfaces that get away entirely from the need to cut into the skull, which will never be without risk.

Many attempts have been made to use variants on EEG (electroencephalographs) to provide an interface to the brain with nothing more than a set of electrodes resting on the scalp. The subject usually wears a plastic cap, which positions electrodes around the skull. As the neurons in the brain fire, tiny electrical charges are generated, which the EEG picks up. Unfortunately, there are so many neurons in the brain that it is currently impractical for an EEG to detect anything other than the average output across millions of different cells. The result, compared with the precision that direct electrodes can provide, is a blurry, limited control that is easily misled by other brain activity. Nonetheless, this kind of EEG helmet, with some kind of transmitting equivalent, shows what might be possible in the future.

The alternative would be to take another route. We tend to think of the brain as purely the lump of matter that resembles an enormous grey walnut in our skull, but the exact line between brain and body isn't always easy to define. The nervous system stretching from head to toe is little more than an extension of the brain. You could say that, for instance, the optic nerve is a system that joins onto the brain, carrying information from the retina of the eye, but equally you could think of it as a part of the brain extending down the optic nerve all the way to the retina.

This is not just playing with words. There is a lot of pre-processing that goes on in the eye before information is sent to the brain. There are many more sensory cells in the eye itself than there are fibres in the optic nerve – the signals from the eye's sensors (rods and cones) are collated before being fired up to the brain – effectively, there's a part of your brain that resides at the back of your eye. This extension of the brain makes it possible for direct connected enhancements – for example, artificial limbs that are mentally controlled – to be wired into the nervous system, rather than directly into the brain itself, providing less risk of infection.

Reframing the universe

The examples we have explored in this chapter have involved creativity and innovation detached from the science that we have used elsewhere in building our universe. However, we should not forget that the insights of Newton, Einstein and the other luminaries we have met along the way were driven by exactly the same kind of creative process. What set Newton aside from most of his contemporaries was his ability to shift his mental frame of reference. Where the general understanding was that heavenly bodies felt no urge to fall in the same way the apple did, Newton's genius was to shift his mental framework to see that the Moon, for instance, was falling in exactly the same way as the apple – it just happened to also be moving sideways at the right speed to keep missing the Earth. Newton reframed the view of gravitation from a local phenomenon to one that spanned the universe.

It would be helpful to take in one more scientific reframing to see the process explicitly in action. This is the move from an Earth-centred to a Sun-centred model of the universe.

It is quite clear, as far as the everyday observer goes, that the Sun rises in the east, crosses the sky in an arc and sets in the west. While it's possible to consider alternative models, the most obvious reason for this is that the Sun is rotating around the Earth. This very reasonable (if incorrect) assumption had not always held, though. Some early Greek philosophers thought, for philosophical rather than scientific reasons, that there should be a fire at the centre of the universe, of which the Sun was just a part seen through a hole. A more considered view was held by Aristarchus, a third-century BC astronomer, who opposed the prevailing opinion by his time that the Earth was at the centre of the universe with everything revolving about it.

The original work by Aristarchus describing his theory is lost, but Archimedes makes a reference to it in his book *The Sand Reckoner*, saying: 'Aristarchus of Samos brought out a book consisting of some hypotheses, in which the premises lead to the result that the universe is many times greater than that now so called. His hypotheses are that the fixed stars and the sun remain unmoved, that the earth revolves about the sun in the circumference of a circle, the sun lying in the middle of the orbit ...'

However, there was little further consideration of the idea that Aristarchus put forward for well over 1,000 years. The interesting thing as far as frames of reference go is that the traditional view that the Sun orbits the Earth is perfectly valid. From the frame of reference of the surface of the Earth, which is the natural one for Earth-dwelling humans to take, the Sun does indeed rotate around the Earth. The problem is that from this viewpoint, not only does the Sun have to move around us every 24 hours, so does the whole of the rest of the universe. So although there is nothing wrong with taking the view from the frame of reference of the Earth, it is a much more complicated view than the 'correct' view that the Earth rotates daily as it travels in its annual orbit around the Sun.

To reach the modern astronomical view, we have to detach ourselves from the frame of reference of the Earth. It isn't enough to move instead to the frame of reference of the surface of the Sun, as that body, like the Earth, is rotating – so once again, we would see the rest of the universe in motion around us. Instead, we adopt a frame of reference that turns with the apparent rotation of the universe, centred on the Sun. Notice again that this is a decision for convenience. There is no absolute frame of reference, nothing that defines being at rest which we can measure all other movements against.

Because of his fame, and the trial he underwent as a result of the way he presented his results, Galileo's is the name often associated with the rotation of the Earth around the Sun, but he was merely passing on the ideas of the earlier Nicolaus Copernicus (actually Mikołaj Kopernik, but we use a Latinised version of his name). Copernicus took some time to settle down into his profession, studying the liberal arts, canon law and medicine, but he certainly had an abiding interest in astronomy, as he is known to have given a lecture on it in Rome in 1500 at the age of 27.

Technically he was a clergyman, paid as a canon of Frombork in Poland, but in practice he mostly worked as a physician and astronomer. He showed doubts about the Earth-centred system in an early book, but his main work on the subject, *De Revolutionibus Orbium Coelestium* (On the Revolution of the Heavenly Spheres) was not printed until he was on his deathbed.

The innovation that Copernicus introduced was doubly a shift of frame – both a literal change of physical frame of reference away from the Earth's surface and also the kind of conceptual shift that this chapter is all about, giving new insights into the workings of the universe. It is also a reminder once more that the frame of reference we use is an arbitrary one. It remains more practical for us, here on the surface of the Earth, to talk of sunrise, say, rather than 'the time at which the rotation of the Earth makes the horizon drop away so that we can first see the Sun'.*

Science goes hand-in-hand with creativity. For a scientist to develop a new theory, he or she has to break away from the old way of thinking – to see what everyone has seen before in a different way. The ability to change frames of reference and study the resulting effects is vital for scientific understanding of the universe around us.

Slippery customer

We like to portray the forward march of science and technology as something driven by highly intelligent men and women, a matter of careful and logical assessment, resulting in incremental

---

* This is also something we ought to remember when pedants moan about there being no such thing as centrifugal force. One of the innovations based on Newton's ideas was to realise that when, for instance, a car is taking a bend at high speed, the force that has to be applied to the car to stop it flying off in a straight line is a centripetal force, towards the centre of the curve it is turning around. Yet inside the car we seem to feel a centrifugal force that flings us towards the outside of the car, quite the opposite effect.

It is commonly said that all that exists is the centripetal force, applied to the car but not its passengers, pulling the car inwards. The passengers continue to move in a straight line until stopped by the far side of the car, so they think they are feeling a centrifugal force that moves them outwards, but this is an illusion. In a sense this description is true. But only if you look at what is happening from an external frame of reference fixed to the Earth's surface. However, as far as the passengers are concerned, the most natural frame of reference is fixed to the car, which from the viewpoint of the passengers isn't moving. From that frame of reference, there genuinely is a centrifugal force – after all, why else would the passengers feel the urge to 'carry on in straight-line motion' when in their frame of reference they aren't moving?

steps in knowledge that take us closer and closer to a better ability to understand the universe we live in. And this certainly happens. But one final mechanism for a shift of reference frame that occurs surprisingly frequently is the fortuitous misunderstanding or accident.

In the case I am going to use as an example, not only are there a couple of changes of frame along the way, there is also a classic innovation myth. The US space agency NASA is always under pressure to justify its existence, and one way it does so is to suggest that many technological developments were a beneficial by-product of its own special requirements. The agency argues, for example, that personal computers would not have developed had it not been for NASA's need for very compact computers on spacecraft. This particular development is borderline, though some inventions like memory foam genuinely were a result of a NASA contract. The development of personal computers was driven more by the availability of cheap, mass-produced chips, rather than the bespoke, highly expensive hardware used by NASA.

Perhaps the most interesting innovations are those that are ascribed to NASA incorrectly. The joker of the piece is the space pen. It is often said that NASA spent millions developing a ball pen that worked in zero gravity, while the Russians simply used pencils. There was indeed a US space pen (which was better than a pencil, as these tend to leave bits of carbon pencil lead floating around in a space capsule), but the pen was developed by a manufacturer as a gimmick, without any request from NASA. More common, though, are the inventions frequently attributed to NASA that actually existed long before the agency, notably Velcro and PTFE, the non-stick substance often sold under the brand name Teflon.

Velcro was first patented in 1948 after Swiss engineer George de Mestral noticed the way that plant burrs stuck to clothing. Here de Mestral had the classic shift of frame from seeing this as an irritating facet of nature to a great opportunity for a product. But PTFE's development depended far more on serendipity. Like Velcro, PTFE was used extensively by NASA both in their spacecraft and their astronauts' suits, but it was an even older development than Velcro, dating back to the 1930s.

In 1938, the American engineer George Plunkett was working at an industrial chemical plant in New Jersey, experimenting with gases that might be used as refrigerants. Plunkett was working with tetrafluoroethylene, a simple molecule with a pair of joined carbon atoms, each of which has two fluorine atoms attached. The gas had to be treated carefully, as in some circumstances it could explode, so it was important to ensure that a cylinder of tetrafluoroethylene was empty before disposing of it. The contents were checked using the simple technique of weighing the cylinder before filling and checking this against its weight in use, to see if it had returned to the empty value.

Plunkett was puzzled by a cylinder that seemed to run out of gas long before it should have – and with a weight that clearly indicated there was something inside. He took the suspect object outside the lab to a blast shield where dangerous materials were manipulated and carefully cut through the cylinder wall. No gas emerged – instead, inside was a white, slippery-feeling plastic deposit. It was already known that ethylene could form long chains or polymers, known as polyethylene (shortened to polythene). It didn't take Plunkett long to realise that the tetrafluoroethylene had reacted under pressure, and, as it later turned out, catalysed by the iron cylinder had polymerised to form polytetrafluoroethylene, or PTFE for short.

On further study, the substance had a slipperiness that was unprecedented in nature. The fluorine atoms make it hydrophobic, repelling water, and its structure left little for other molecules to cling on to – even the electrostatic van der Waals forces that enable a gecko to climb smooth walls fail when faced with PTFE.

The DuPont subsidiary where Plunkett worked soon patented the substance, giving it the tradename Teflon, inspired by the recently patented Nylon. Developed in an industrial setting, the obvious applications of PTFE were ensuring that valves and joints were well sealed – and it's still used that way by plumbers today. However, the most visible use of the substance came about with an example of someone seeing an opportunity to change the frame of reference because they weren't knowledgeable enough to know that it wasn't possible.

In the early 1950s, French engineer Marc Grégoire had got some PTFE tape to use on the joints in his fishing tackle. Grégoire's wife immediately saw PTFE in a very different light from her husband. She thought that this slippery substance would surely prevent food sticking to her pans if they could be coated with it. As an engineer, Grégoire had serious doubts about the practicality of this, especially as there was insufficient evidence of PTFE's response to the kind of heat it would suffer in a pan. But he thought it worth a try.

In practice the temperature was not a problem – but there was still an almost laughably predictable issue. Grégoire couldn't get the non-stick material to stick to a pan. In the end, he got round this by a small shift of reference frame – he changed the pan to accommodate the plastic film. Pans were traditionally very smooth on the inside to reduce the tendency of food to stick. But for this purpose he needed a surface to encourage sticking. Grégoire etched the surface of the pan with acid and found that when PTFE powder was sprinkled over the surface and heated, it used the pits to bond to the material. By 1956, his small factory was producing the first non-stick pans, trading on the Teflon name by making his brand Tefal.

A non-stick substance may not represent the same kind of leap forward in human ability that, say, writing provided, but it illustrates well the varied mechanisms by which a shift of frame can take place.

The exceptional ape

We humans have gone way beyond anything that could be possible biologically or that we could realistically envisage getting to via a biological route. It took birds millions of years to evolve the ability to fly. We can go from our natural state of never having left terra firma to sailing through the clouds in hours by buying an airline ticket.

Going even further, we now have the ability that no other animal has ever matched, of travelling into space. As yet this is an extremely tentative venture. All space travel that human beings have ever undertaken has occurred during the lifetime of a single living generation. It is hard to believe, though, that if the human race still exists in a few hundred years' time, and if it hasn't managed to destroy civilisation, that space travel will not have become relatively commonplace. And that, in principle at least, brings with it the chance to take the next step of development – to see human life spreading beyond the Earth. A clear marker that humans are very different from other life on Earth, something that some biologists struggle to accept.

Evolutionary biologists are quick to point out that human beings do not mark some kind of evolutionary pinnacle, which is a meaningless concept since evolution has no end point. Nor has *Homo sapiens* emerged from natural selection as a target that was identified to be 'better' than our predecessors, following a whole chain of improvements that brought us to our present form from ape-like ancestors, as is often illustrated in the traditional 'evolutionary chain of man' illustration. There is nothing directed about evolution. However, it is entirely possible to understand this blind process and still be of the opinion that there is something special about humanity.

Figure 10: The fictional 'evolutionary chain of man'.

Many biological theorists dislike the idea that humans are in some sense exceptional, going so far as to use the term 'exceptionalism' as a kind of insult. And yet for the objective observer it is difficult *not* see humans as extraordinary among life forms on Earth, in terms of our understanding of the nature of the universe, our ability to communicate that understanding, and our ability to change our environment and to enhance ourselves far beyond our biological limits.

The theorists argue that this does not matter – because, for instance, a badger doesn't need any of the results of our intelligence and civilisation to survive. If we avoid anthropomorphism, the suggestion is, making a comparison purely in the light of survival of a species, our abilities may not be as exceptional as we think. A good example of this concern that we shouldn't give humanity any special position comes from palaeontologist and journal editor Henry Gee, commenting on the TV series *Human Universe*, written and narrated by TV physicist Brian Cox. Gee writes:

> Cox speaks, with the prerequisite Bronowskian awe and reverence, of our uniqueness as a species, that we are the only species capable of doing the things we do, by virtue of attributes such as language and writing. Cox turns his boyishly unfocused gaze of general wonderment from the heavens to the depths of antiquity, the growth of societies and trade and how writing pulled this all together.
>
> It's this – the assertion of the uniqueness that makes us special – that really gets up my nose, because it's a tautology and therefore meaningless. Giraffes are unique at doing what they do. So are bumble-bees, quokkas, binturongs, bougainvillea, begonias and bandicoots. Each species is unique by virtue of its own attributes – that's rather the point of being a species – and human beings are just one species among many. To posit humans as something extra-special in some qualitative way is called human exceptionalism, and this is invariably coloured by subjectivity. Of course we think we're special, because it's we who are awarding the prizes.

But this argument is limited. Our minds, and the things that they have enabled us to do through our creativity, have set us apart in a way that *is* unique among known species, in both the scope and impact of what we have been able to achieve. In *The Ascent of Man*, Jacob Bronowski states:

> Man is a singular creature. He has a set of gifts which make him unique among the animals: so that, unlike them, he is not a figure in the landscape – he is a shaper of the landscape.

Bronowski argued that our lack of specific fit to an environment, which may seem initially a huge disadvantage, should be seen instead as a valuable ability to fit *any* environment. He considered that our rather poor 'survival toolkit' has thankfully proved flexible enough to cope with many different possible challenges; it gives us a unique ability to not only fit into an environmental niche, but to transform the environment. We act on our environment not in the unconscious fashion of a grazing herd, or the extremely localised modification of building a nest or burrow, but by radically modifying the world around us to fit our needs.

Even if we accept Gee's argument that plenty of other species don't need to make use of the kind of creativity we display, this is only true as long as their environment continues to support them unchanged. If the environment undergoes drastic change, without creativity the species won't survive. We only have to look at the disappearance of a biological group that proved successful for many millions of years – the dinosaurs.

Although there are still plenty of living organisms that many biologists now accept to be of the group dinosauria – the birds – the non-avian dinosaurs appear to have been wiped out by drastic environmental changes, probably caused by the combination of an asteroid collision and volcanic activity. We are the only species on Earth that has, in principle at least, developed the potential to survive such drastic changes to our planet, either by modifying our environment or, in the worst case, by leaving our home world and voyaging to another.

◇◇◇◇◇

With human creativity and innovation providing the last contribution to our model universe, we now have the opportunity to take a step back and see how the difficulty of finding a shared frame of reference can mean that scientists and the public struggle to comprehend each other. While scientists have put together our best models of the universe, they still find it hard to produce a clear and effective view of science itself that the public can understand and support. And without a shared viewpoint it can seem impossible to get a clear picture of our place in the universe.

9 Fundamental Relations

◇◇

Even now, over 2,350 years after Plato's death, we are haunted by his vision of the existence of an overriding absolute, by comparison with which all we know and do is but a shadow. The need for absolutes seems to be present in many of the oldest human approaches to understanding the universe. Although some religions have been relativistic, with gods that were nothing more than super-powerful versions of humans, many make the central deity or deities absolute in their power and reach, putting them outside the sway of relativity. But despite the enthusiasm that scientific atheists have for attacking religion, this isn't a significant issue for the topic of this book. Science really has not got a lot to say about religious claims.

The fact remains, though, that most of us try to force the world, and science, into a structure that is based as much as possible on ideals and absolutes. We want science to be black and white. We want science to provide us with unqualified facts. To turn its spotlight onto 'the absolute truth'. We get frustrated when scientists hedge their statements with provisos. So strong is this aversion that it is very tempting for scientists, especially when faced with the soundbite world of broadcast science where there just isn't time to explain everything, to drop the provisos and make statements that sound as if they too believe that their theories and models are fixed and absolute truths.

We (and the scientists and broadcasters) need to remember that science is always provisional. All science can ever do is to provide theories that fit the current data as closely as possible and that may need to be thrown away tomorrow if new data comes along. A model – the form that is used by much of science, from the widescreen drama of the big bang to the minute detail of quantum theory – can only ever provide a view through a particular intellectual frame of reference. This is why we can quite happily say in apparent self-contradiction that light is like a wave, and like a particle ... and, for that matter, like a disturbance in a quantum field. It depends which reference frame you use.

Where ideals have crept into the realm of science, we have seen science become corrupted by political values – as when the Nazi and Soviet regimes imposed bad science on their academics to make any discoveries fit with political ideals. Similarly, there are examples of studies where findings have to be ignored or cherry-picked, because they could be considered politically incorrect or otherwise unacceptable. Politicians usually struggle to think in anything close to a scientific manner (not helped by the way that the vast majority of career politicians come from a humanities background). To a politician, obsessed with absolutes, a U-turn is a disgrace, to be avoided even at the cost of doing something disastrous. To a scientist, making a U-turn is the natural result of fascinating new insights – something that should be welcomed, and applauded, when appropriate.

Admittedly, it can take a scientist a while to come round to a change of view, as was the case with American physicist Robert A. Millikan. His greatest claim to fame is probably determining the charge on the electron, but he also contributed significant evidence to support quantum theory – while trying to disprove it. Millikan was convinced that Einstein was wrong when he considered light to be packaged up as particles in his explanation of the photoelectric effect (this was the 1905 paper that won Einstein the Nobel Prize). Millikan *knew* that light was a wave. He had been taught this since his earliest training in physics, and there was plenty of evidence to support it.

In order to disprove Einstein, Millikan ran a series of experiments, recording data from the photoelectric effect in far more detail than had been possible before – and confirmed Einstein's theory time after time. Millikan was human, and resented being wrong – but his experimental evidence was overwhelming and others immediately applauded its significance. In many other cases, the discovery of an unexpected result that requires a change of thinking is regarded as an immediate win by a scientist. Simply confirming the status quo does not add anything to our understanding of the universe. But finding something new, especially something that contradicts earlier theories, and so could open up whole new pathways, is exciting indeed.

One of the problems we have in getting a handle on the nature of science is that rationality, the best tool we have to bring to bear, is not an absolute mechanism itself, but something that has a relativistic component. The nature of just what is rational is interestingly highlighted by the psychologist's favourite game, the ultimatum game, in which one player divides an amount of money between the two players involved, and the second player then decides whether to accept this split, or to reject it, in which case neither gets anything.

An absolutist view of rationality observes that the amount of money awarded to the second player in the game is irrelevant. If this player is offered anything, they are getting money for nothing: that absolute value should be the deciding factor. According to this view, he or she should inevitably take any cash on offer. But human rationality takes the *relative* amount they receive to be more important than the absolute. As a result, most people will turn down free money if they are not offered a high enough percentage because they feel they are being mistreated by the person who decides to keep the lion's share. Only with large sums, where even a small percentage produces a life-changing amount of cash, do absolutes dominate. A true view of human rationality has to take into account circumstances where a relative viewpoint is important.

Games are all very well, but they aren't reality. However, we do have actual examples of the way that using the correct frame of reference can be essential for rationality, even if it ends up being ignored by politicians. Consider the summer time experiment of 1968. The UK government decided to try out keeping the country's clocks on British Summer Time (BST) all the year round, rather than switching them back an hour to Greenwich Mean Time (GMT) over the winter months. The experiment was a huge success. There were around 2,500 fewer casualties on the roads, with several hundred lives likely to have been saved. So what did the government do? Even before the experiment had finished, they announced that the UK would go back to alternating between BST and GMT. And we did. Which is what the country's clocks have continued to do ever since.

To cancel a highly successful trial was not rational – but it fits perfectly with the politics of the heart. The argument was that even though the total deaths and injuries on the road went down, the number of accidents in the *mornings* (when it was dark for longer than it otherwise would have been) went up. And to the politicians, specific individuals, voters who could appear in the media berating the government for causing the death of their child on their way to school, were much more important than the far greater number of unidentified people whose lives had been saved. The lives saved were just a percentage, a relative statistic that made sense to science, but they were not the absolute presence of talking heads on the TV, which carried far more weight in politics. Heart won over head, and in the years since the trial we have probably lost around 20,000 lives in the UK unnecessarily. The rational statistics lost out to the impact of specific mourning relatives.

As a larger-scale example, there is good evidence that our happiness and well-being are more dependent on our relative position compared to others than they are on absolute values of wealth, earnings, possessions and so forth. We are less concerned about how much we have in the bank than with how wide the split is between the ends of the spectrum and with our relative position on that spectrum. We see ourselves, and the world around us, through relativistic eyes.

This comes through strongly in the evidence presented in the book *The Spirit Level*, which uses statistics to demonstrate the corrosive impact of inequalities on quality of life. Authors Kate Pickett and Richard Wilkinson's research suggests that 'ill health, lack of community life, violence, drugs, obesity, long working hours, big prison populations' are all more likely to occur in a less equal society. It is only by taking a relativistic view of social and financial positions that we can make a better attempt at creating a society that works. Yet few political systems seem capable of accepting this message.

Super structures

When we try to establish the relationship between humanity and the universe, as we saw in the previous chapter, some biologists struggle hard to avoid any suggestion that human beings are special. This urge seems to combine both psychological and religious factors. From the psychological viewpoint, the aim is to overcome a natural self-centred tendency for humans to think of ourselves as the most important things in existence. Moving away from such a view is clearly important for scientific objectivity. What human beings have done is incontrovertibly unusual and worthy of note, but we must not confuse that with thinking that there is something inherently unique about the human race. However, so strong is the urge to counter this bias that science often overcompensates.

Similarly, the scientific community has demonstrated a repeated urge to show that humans don't have some sort of special God-given status. Until recently there was a tendency, particularly in cultures based on the Jewish, Christian and Muslim faiths, to assume that because of the biblical emphasis on humanity's dominion on Earth and a special relationship that the religions claim we have with God, that human beings must accordingly have a unique position in the universe.

The result is an overcompensation that can bias scientific decisions, making them not as a result of actual scientific evidence, but with the intention of ensuring that a belief system does not triumph – and the trouble is that the scientists who do this are themselves then shifting from using the scientific method to employing a belief system. The most commonly cited example of this is the approach taken by the aggressively atheist followers of Richard Dawkins, but a more interesting belief-based issue is proving a challenge in astronomy.

Astronomers and cosmologists have something of a history of this kind of bias. The great Fred Hoyle, for example, was vehemently opposed to the big bang theory because it fitted well with the possibility that the universe had been created. More recently, as astronomers have explored the depths and detail of the universe with increasingly powerful telescopes, they have discovered something that runs counter to widely held scientific belief: large-scale structures. There is, for instance, a huge gap in the observable universe around 2 billion light years across which is, in terms of the typical density of galaxies everywhere else, pretty well empty. In a different location in the universe, by contrast, there is a vast string of 73 quasars.

Quasars are remarkable light sources, each pumping out as much electromagnetic energy as a whole galaxy. They are believed to be the radiation from material that is plunging into supermassive black holes at the heart of young galaxies. This string of quasars forms an immense structure that is 4 billion light years in length. Similarly, there are intense bursts of gamma ray energy that form a ring taking in a good 6 per cent of the visible universe. As far as cosmologists are concerned, none of these structures should exist.

The cosmologists do not make this claim based on good science. Instead, their response is a belief, based on something given the impressive-sounding name of the 'cosmological principle', which amounts to saying that the universe should be the same wherever you look. It's not right that there should be large-scale structures, the cosmologists argue, because there shouldn't be any special places in the universe.

At first sight, the cosmological principle seems to be the

most bizarre suggestion in all of science. If there is one thing that even the most amateur of astronomers can say, it's that, looking out into space, things are very different depending on where you look. It's hard to get a bigger contrast than that between a stretch of black, empty space and the huge, brilliantly glowing nuclear reactor of a star – let alone comparing a near-void like the 2 billion light-year gap with the vast structures of the billions of galaxies we know are out there. Just look at the spiral complexity of our nearish neighbour the Andromeda galaxy (or, for that matter, our own Milky Way galaxy). That is clearly not the same as empty space.

However, supporters of the cosmological principle regard these kinds of structures as trivial. Yes, of course the universe is not *really* uniform as far as fine detail goes, they would argue. But in universal terms, stars, and even galaxies, are extremely fine detail. We are being parochial, seeing things from the scale of tiny human beings on a teensy planet, for whom a star or a galaxy is big, but really such structures are insignificant. If you look on the large scale, the cosmologists argue, ordered regions like galaxies will average out, leaving no significant difference between parts of the universe. It's all much of a muchness. There will be no huge gaps with nothing much in them. There will be no vast structures, spanning sizeable chunks of the universe. And yet vast structures and gaps are precisely what are being observed. The universe is not uniform.

The cosmological principle is based on two foundations – the Copernican principle and the assumption that the universe is isotropic. The first states that there is nothing special about the location of the Earth and the second that the universe appears the same wherever we look. The observations being made of large-scale structures seem to call the isotropic assumption into question. As for the Copernican principle, the basic concept is correct, but the way it has been extended is doubtful.

Cosmologists argue that there is no reason to give Earth a special place in the universe. We once put it at the centre of everything, but we now know that this was a mistake. We are just a backwater planet. And purely from the point of view of the location of the Earth, the Copernican principle has proved to be absolutely true. But all too often it is used in a similar fashion to the biological 'nothing special' argument, and there it becomes more of a belief system. Cosmologists argue that because the Earth's location can't be considered special, then the inhabitants of the Earth must also be nothing special, once again minimising the significance of human achievement.

Like the idea of unchanging natural laws and universal constants we met when first exploring the concept of space in Chapter 2, the cosmological principle is an absolutist viewpoint that has been applied for convenience. Without the principle, it becomes much harder to apply physical theories to the universe as a whole. In fact, the extension of general relativity to go beyond a simple explanation of planetary orbits and why things fall, to be able to make grand, if highly simplified, models of the universe – an occupation that began as soon as Einstein's results were available and that fascinates theoretical physicists to this day – depends on the assumption that the cosmological principle is true. Without the principle, it would be pretty well impossible to apply general relativity on the scale of the universe, as the equations would have to be applied to a structure of varying density, rather than to something that can be assumed to be uniform. But there is no science to support the existence of the cosmological principle.

What is scary, given the pragmatic way that the overall

principle was constructed, is that, rather than contemplate the most likely outcome that the cosmological principle is a convenience, a hope and belief that doesn't really work out in reality, cosmologists are prepared to go to remarkable lengths to try to brush the contrary evidence under the carpet. One hypothesis, for instance, is that the large-scale structures we see in the universe don't exist at all. It has been suggested with a straight face that the structures are 'just' the side-effect of projections from the extra dimensions required for some attempts to merge quantum theory and general relativity, like M-theory (an extension of string theory). It's hard not to see such an argument as an act of desperation. It's a bit like holding a reasonable-feeling assumption like 'all swans are white', then, when you are presented with a black swan, rather than realising that the assumption was an over-simplification, suggesting that the apparent black nature of the swan is an optical illusion caused by an inter-dimensional leakage of blackness from outside our universe's brane.

Not all explanations for the apparent large-scale structures are quite so difficult to swallow. They could, for instance, simply be random occurrences that require no cause. Randomness is a hard thing even for working scientists to get their heads around. Our natural reaction to discovering that something is distributed randomly is to assume that it will be roughly evenly spread out – but in reality, clusters and gaps happen far more in random behaviour than we tend to expect. If, for instance, there is a cluster of cancer cases in one location, the natural assumption is that there is a collective cause, and we will look for something to blame, even if the chances are that such a cluster is an expected statistical fluke.

If we look at a simpler example, the way that random

distributions will naturally have clusters and gaps feels a more reasonable outcome. For example, if we drop a boxful of ball bearings on the floor, we would not expect the balls to come to rest in a nice even distribution. We would expect there to be areas with clusters of balls together and areas with very few or no balls at all. If they were evenly spread, it would seem that there had to be some kind of organising force at work – perhaps a series of magnets under the floor.

It is possible, then, that the apparent structure of the 73 quasars is just such a random clustering. How likely this is depends on the size of the universe – which is a question for which we just don't have an answer. If the whole universe were vastly larger than the part that we can see, which is about 90 billion light years across – perhaps even infinite – then it's entirely possible that a structure like the string of quasars would occur now and again and it just happened to fall in our part of the universe. This would explain why such structures might naturally occur without a cause. But unfortunately the observation still prevents the cosmological principle from holding on the scale of the visible universe and still presents cosmologists with problems.

Fine-tuning

While Victorian scientists gloried in the wonders of human ability, tending, without doubt, to overrate our significance as a race, modern science has turned the tide to such an extent that a kind of scientific political correctness has set in. Science can now go too far to avoid a special place for us. *Homo sapiens* is treated to a kind of inverted snobbery, where every effort is exerted to ensure that nothing special is claimed about our existence. This also means there has to be nothing special about the Earth, our solar system, the Milky Way or any particular part of the universe seen on an appropriate scale.

This 'specialness denial' not only happens when the Copernican principle is applied; it also steers the reaction of some cosmologists to the way that our universe itself seems to be fine-tuned in many different kinds of ways. Many parameters and constants would only have to be slightly different and life (or planets, or stars, or galaxies) would never have come into being.

The problem that cosmologists and astrophysicists have here, which is often mistakenly picked up by those trying to find evidence for a creator, is that this fine-tuning seems just right for the existence of stable physical structures and for the eventual development of life. Each of these builds on the others. Without the ability for galaxies and stars to form we wouldn't have planets. Without planets in a stable solar system at the right kind of distance from the right kind of stars, life is unlikely to have started. Without the right chemical reactions and right materials, again life seems a non-starter.

What's more, our universe seems at a general level to be far too geometrically flat, consistent and relatively empty for the predictions of physical models. It is an oddity in this respect. Although it is one possible configuration, the vast majority of ways the universe could be structured would not be so flat, smooth and relatively empty. To quote cosmologist Lee Smolin:

> If we reached into a hat filled with pieces of paper, each with the specifications of a possible universe written on it, it is exceedingly unlikely that we would get a universe anything like ours in one pick – or even a billion.

We live in a special kind of universe, it seems, and as this runs counter to the large-scale application of the Copernican principle, some cosmologists assume that there has to be a cause for the outcome.

This apparent unlikeliness leads to some dubious logic that suggests the only way we should be living in such a special universe is if there is a multiverse, a super-universe that contains many different constituent universes, allowing for a vast range of possible structures and laws and universal constants. In such a multiverse, it is suggested that there is no need for our universe to be special because with all the different possible universes, it is inevitable that some of them will have the appropriate fine-tuning that would enable life to flourish.

This could be described as the lottery argument. Lottery winners exist, even though it is incredibly unlikely that any specific individual will win. Because there are vast numbers of players, many of whom have different tickets, it becomes possible to have winners week after week, even though the chance of winning with any particular ticket is millions to one against, as we saw in Chapter 7. Similarly, the lottery argument says, with vast numbers of universes available, many of them with different values for universal constants and different natural laws, it is possible for our universe to be (at least) one among the multitude. The supporters of the existence of the multiverse then invoke the so-called weak anthropic principle, which says that we can only be here to observe our universe if it supports life, so we are bound to find ourselves living in one of those improbable universes where life can exist. It's very unlikely that our universe should be set up this way, but because there will be occasional universes in the multiverse that have these parameters, and these are the ones most likely to support life, inevitably, we will find ourselves in such a universe.

The problem with this kind of hypothesis as an explanation for what we experience is that it isn't really a scientific approach. The 'one unlikely universe in a mix-and-match multiverse hypothesis' makes no predictions that can be observed and checked (other than the existence of the universe, which is a circular argument, as it's where the idea came from in the first place). And it gives us no mechanism to explain why we have the particular set of laws and constants that exist in our universe.

Apart from anything else, even if there were a multiverse, this does not give us an explanation for why different universes in that non-communicating framework should have different universal constants and natural laws. For the constants and laws to be distributed across the universes there has to be some overview or communication between them. As Smolin puts it:

> As attractive as the idea may seem, it is basically a
> sleight of hand, which converts an explanatory failure
> into an apparent explanatory success. The success is
> empty because anything that might be observed about
> our universe could be explained as something that
> must, by chance, happen somewhere in the multiverse.

In reality, this kind of multiverse hypothesis is science fiction in search of a reality. It's a bit of fun, and it gives the more speculative cosmologists something to write endless books about, but it has no scientific value. It isn't even a necessary hypothesis. The intention of devising this mix-and-match multiverse is to explain how an unlikely universe happens to exist, but in an infinite range of possibilities, *all* possible universes are infinitely unlikely. So if we take the starting point that there is just one universe, there really is no reason for it not to be ours.

Imagine you go to a busy motorway and take note of the number plate of the first car that passes under a bridge: it happens to be BC15 GDS. It is extremely unlikely that this particular number plate out of the millions that are on the road should happen to have passed under your bridge. And yet it did. There had to be some next car – it just happened to be this one. Similarly, if there is just one universe, it is very unlikely it will have any particular combination of laws and constants and so forth. But it has to have some values – and we've got what we've got.

Once we have the single universe that just happens to be one with useful parameters for life, we can once again apply the weak anthropic principle and say that we would not be able to think about this universe and discuss it if it hadn't happened to be the particular kind of universe it is. The chances are very unlikely that the universe would take this form (or any other specific form), but once it had, life was possible, and had it not taken this form there would have been no life, no consideration of the matter and no book for a non-existent audience to read.

Those who insist on the multiverse are convinced there must be an absolute external reason for the particular form our universe takes, rather in the same way that in earlier times human beings needed an absolute external deity to establish *why* creation took the form it did. They struggle to accept that we exist in the universe we do. Yet discarding the absolutist viewpoint gives us the most meaningful way to explore our situation. If we can accept that we don't need a multiverse to make it possible for our universe to exist, then we can start to ask the more interesting questions that the multiverse approach ignores. Why are the natural laws and constants the way they are? With a multiverse, there is no 'why' because we just happen to be in the one of many universes that has these values. Similarly, without the multiverse we can consider the interesting possibility that these constants and laws can change over time. Moving away from the absolutist idea that our particular universe comes with a fixed set of laws and constants that can never change with time is a valuable start in trying to find out if there is any way to determine *why* those values hold in the universe.

No closed systems

As Lee Smolin has pointed out, a common approach to physics is what is sometimes called the Newtonian paradigm. This involves taking a closed system like an experiment in the lab, or a star (or for that matter a universe), and applying a set of starting conditions and natural laws. As a result of the starting conditions and the laws you can then predict what will happen in the future. Even where probability creeps in, as it does in quantum theory, this approach still stands. But the problem with this method is that it can't be used to discover why the laws and initial conditions are there in the first place. To do that would require a different viewpoint.

One assumption that it may help to challenge is the 'closed system' one. In science we almost always assume that we can put an experiment in a box and isolate it from its surroundings. But in the real world this hardly ever happens. This is one of the reasons that it is so difficult to make accurate studies, for instance, of how different foods or lifestyle choices influence health. In the real world we can't put people in a box and control every other aspect of their lives apart from the one being studied. A person isn't a closed system, but has many (many) interactions with the world around her, all potentially producing changes to health and well-being.

Similarly, in physics, there are no totally closed systems. Gravity, for example, is unstoppable. There is no box in which we can put an object to prevent it being influenced by the gravitation from outside that box. The ability to produce an antigravity shield, like the material cavorite in the H.G. Wells novel *The First Men in the Moon*, bumps up against one of the more robust laws of physics: conservation of energy. If we could shield against gravity it would be easy to break this and produce a perpetual motion machine.

Imagine a traditional waterwheel, but each paddle is painted on one side with cavorite, so that side of the paddle is shielded from the Earth's gravitational pull. The other side would be pulled towards the Earth, producing the same kind of rotation the wheel experiences with the force of water on the paddles, but simply driven by gravity. Connect that wheel to a generator and you've got limitless free electricity. It's not going to happen.

Without our cavorite shield it is impossible to consider the system made up of the Earth and the Moon without bringing in, at the very least, the influences of the Sun and Jupiter. Technically every other body in the universe that has been around long enough for its gravitational influence to reach us makes its own tiny contribution. The Newtonian paradigm of establishing starting conditions and laws of a closed system is a handy simplification, but it is never an accurate description of reality.

It might seem that when we consider the universe as a whole, assuming it is finite and so can be considered closed, that we have a true closed system. In a sense this is true, but it is still different from the kind of system beloved of physicists. This is because we – and the scientists – are inside that system. The assumption is usually that the experimenter is separate from the experiment. Quantum theory challenges this possibility, but cosmology has to totally ignore it. By definition, we are part of the universe and there is no way that we can observe the system from the outside. If the universe is infinite – which is possible – or even finite but unbounded, we have the greater challenge still that there is no possible way to close it off from what amount to external influences (in the finite but unbounded case, strictly speaking the influences are internal but self-acting).

All this means that any attempt to understand why the natural laws are the way they are needs a different way of thinking from that used in the Newtonian viewpoint. Whether this is even possible for anything other than theology or philosophy is not clear. But if it is to be possible we need to let go of the urge to dismiss the remarkable nature of existence for both the universe and humanity: to accept that we are special and that this is something we should celebrate and explore, rather than try to deny.

Smolin suggests, in collaboration with philosopher Mangabeira Unger, that there are three principles required to produce a theory that explains the natural laws and the initial conditions of the universe. These are that there is one universe, that time is real with the laws of nature changing over time, and that mathematics is not about an ideal reality but a description of the real world.

These are interesting suggestions. While there can be no more evidence that there is one universe than that there is a multiverse, it seems a sensible starting point, if only by the application of Occam's razor. This famous principle of the English friar William of Ockham (or Occam) is generally paraphrased as 'Choose the hypothesis with the fewest assumptions', though his original formulation was something like 'Never choose plurality unless it is necessary'. This fits rather nicely with not positing a multiverse unless there is specific evidence for its existence, rather than making use of the pseudo-science argument: 'The particular form our universe takes is unlikely, so it must be an unusual one of many.'

The second of Smolin's suggestions addresses a particular problem he has with the assertion from many physicists, not that time literally doesn't exist, but rather that the 'flow' of time is largely irrelevant to many physical processes. Certainly it would be impossible to suggest that natural laws were timeless if they did change during the existence of the universe – and changing laws are also a useful precept if there is to be an explanation for why a particular set apply. An unchanging set allows no mechanism during the existence of the universe for the laws to get to their present state. If laws cannot change, they have to be imposed at the start by some kind of mechanism external to the universe, whether it is god (literal or metaphorical) or an earlier universe which gave rise to this one.

There is a moving episode in James Blish's science fiction novels, the *Cities in Flight* series. This quartet of books begins as an adventure story with the remarkably imaginative concept of 'spindizzies' – mechanisms by which whole cities can be lifted off a dying Earth and taken into space to undergo faster-than-light travel. However, towards the end of the series Blish becomes surprisingly philosophical for 1950s science fiction and dreams up an end-of-the-universe scenario.

Rather than have some kind of ultimate get-out clause, a *deus ex machina* to save the protagonists, the universe does genuinely end as the book draws to a close. But the main characters have found a particular location where, if an object could survive into the formation of a new universe for a tiny fraction of a second, that object would define the laws and nature of the new universe. With special technology, each of the main protagonists becomes the seed of his or her own new universe. We might not live in a world where such technology exists, but Blish was, at least, unusual in thinking about where those laws might have come from in the first place.

The last of Smolin and Unger's principles is perhaps the strangest. It isn't totally clear why they regard it as so essential for mathematics to be the description of the real universe if we are to attempt to explain the origins of the natural laws. It is certainly true that the very early foundation of mathematics in counting was tied to actual physical objects, but maths very quickly took on a life of its own. What we now find is that mathematics operates independently of the universe, but that we can often make useful deductions from mathematical models that parallel the universe. Mathematics has ties to reality, which might be enough for Smolin and Unger's exercise, but it can't be truly said to be limited to our universe.

What certainly does seem to be the case – and some physicists have specifically worried about this – is that there has been a tendency to derive modern physics from mathematics more than from observation and experiment. Perhaps what Smolin and Unger had in mind was that without this mathematical core to reality, it would be very difficult to use mathematics to make any suggestions for reasons that particular laws and constants apply. However, this seems an unnecessary restriction.

If mathematics is, as seems more likely, an arbitrary tool that can often be usefully employed to model reality, there is no reason why it can't be equally of use in the pursuit of the origin of the natural laws. (Surely, if maths were totally tied to reality then it would be dependent on those laws and would be unable to approach them from the outside?) We just have to always be aware that when using mathematics to provide predictions about the universe, we are dealing with models, not a meaningful description of how reality actually works. Mathematics is not an absolute measure of the natural world, but a relativistic parallel.

Framing the universe

It is only by letting go of many things that we have a natural tendency to regard as absolutes and accepting the relativistic view that we can understand what science is telling us about the universe and about ourselves. And it is only by understanding relativity that we can counter the ways in which it is misused in a hand-waving sense to suggest that a vague statement like 'Everything is relative' allows us to make the leap to 'Every theory has equal value', which is certainly not what scientific relativity is all about. It is important for us all that we understand the true significance of relativity and can truly assess human achievement to date.

The aim here is not to regain the nineteenth-century picture of humanity as the pinnacle of life, nor to regress to the more ancient picture that put humans and the Earth at the centre of the universe. We know that we are a very small part of a huge universe. We know that we are just one stage in an evolutionary process that can see life go in very different directions in the future. Evolution, after all, is not directed towards some ultimate, absolute goal. It must always function relative to its environment and our competition – and these will change, either incrementally or drastically with time.

However, accepting all this does not mean that we can't also think of ourselves as special. Those who depend on the Copernican principle are stuck in an old-fashioned viewpoint that requires an absolute view of the vastness of the universe and our tiny place in it as just one mammal out of so many species on Earth. In Kate Pickett and Richard Wilkinson's terms, this is like rating human beings based on the absolute rather than the relative amount we have in the bank. But a scientific view is not an absolutist one. Whether dealing with physics or biology, relativity has to come into play. When positioning human beings we have to make use of the frames of reference we have available.

As we have seen, there is reasonable evidence that life, particularly complex life, is likely to be rare in the universe – that we are a rarity in existing at all. And there is no doubt that the way we have used creativity to build on top of our biological capabilities, enhancing the basic human, sets us well apart from all other species on Earth. What we have achieved in science and technology in just a couple of thousand years is genuinely remarkable. We should celebrate human achievements, which – relative to all other life forms we are aware of – are outstanding.

In the end, the relationship of science and society is critical to sustain our technologically based lifestyle and indeed the future of humanity. And relativity is an essential tool in making sure that science continues to use appropriate reference frames to understand our relationship to other living things and our potential to extend beyond the limitations of a single planet, helping us to assess the risks and possibilities that science itself will bring us for the future.

As a signpost for the significance of our capabilities – and a constantly recurring theme as we have built a universe in this book – it is essential that we put relativity at the heart of science.

Appendix:
The Special Theory of Relativity for Beginners

◇◇◇

There is no need to read this extra material to enjoy the book, but if you were intrigued by the suggestion that, whereas general relativity was a stretch mathematically even for Einstein, the maths in special relativity could be followed by a high school student, I wanted to provide the opportunity for you to see just how the fixed speed of light results in the remarkable time-bending results of Einstein's special theory of relativity.

There will be equations. If you have problems with these as a hangover from school days, bear in mind that an equation is just a shorthand that makes it easier to see and manipulate mathematical entities – or in our case, to model the behaviour of physical things. Each of those scarily compact letters in an equation merely stands in for something straightforward. Some of these will be constants – in effect just useful numbers. So by using c, for instance, in an equation we are saved from having to type 'the speed of light in a vacuum' or 299,792,458.

Equally, one of the letters could stand in for a variable. A variable is harder to get your head around than a constant, but is extremely useful because it's like a container into which you can put whatever is most relevant to you. So, for instance, we will often come across v – short for velocity. The useful thing about a variable like v is that we can pop in it whatever value is relevant to the thing we are studying. So if we're looking at a spaceship travelling at 100,000 kilometres per second, then we can just pop 100,000 into v and we're away.

The only proviso here is about units – a unit being something like 'kilometres per second' which describes the scale of the numbers being used. It's fine to use any unit you like that fits your requirement. In the case of velocity it happens to be a unit of distance per unit of time. So I could equally well use feet per year. Or, for that matter, your height (whatever it happens to be) per millennium. The only problem is that once I've decided on a distance unit, I need to use the same one for everything in the equation, or else I'm going to have problems.

You can see why this is the case by thinking about speed limits. If the limit on a road is, say, 60, meaning '60 miles per hour', it's no use saying 'I was only doing 2', meaning '2 miles a minute'. You need to use the same units as the limit does – in this case, making your speed a more fine-inducing 120 miles per hour.

To avoid having to check the units being used all the time, there are standard units in the common scientific system known as MKS, which stands for metres, kilograms and seconds. So we would usually expect scientists* to measure velocity in metres per

---

* This is with the exception of astronomers, who never entirely got the hang of fitting in with everyone else. Not only do they not measure distances in metres, but they have two different units of distance, the light year and the parsec. And they call everything not hydrogen or helium (carbon or oxygen, for instance) a metal. Even scientists think astronomers are a little strange.

second, making the speed of light just under 300,000,000 metres per second.

So now that, hopefully, equations hold no real fear for us, let's try out the most famous equation in existence, one that has already appeared in this book:

$$E = mc^2$$

This has all the basic aspects of an equation in a simple, digestible form. It's an equation because of that equals sign. That tells us that the thing on the left (E) has the same value as the thing on the right (mc^2) – they are equal, hence '*equation*'. We've already met c as a constant: it's the speed of light, but this time it is squared – multiplied by itself. And then we've two variables, energy (E) and mass (m). And the neat thing is, either of these can be given any value we like. We then just plug in the numbers and we get the other one. If we know energy, we can work out the equivalent mass. Know the mass and we have the available energy. It's as painless as that.

A final useful extra when dealing with the equations of special relativity is that we need to differentiate between, for instance, the time on the Earth and the time on a spaceship. It is traditional in books on relativity to indicate the two different 'frames of reference' by having one set of variables that are normal (e.g. v, t) and one set that have a little blip attached – v' and t'. I think it is much clearer to be explicit with the labels and call them, for instance, t_{Earth} and t_{Ship}. I suspect the reason that academics tend not to do this is partly because they have to deal with much more complex equations than we will be handling, and partly because the 'dash' notation started off on a blackboard, where it would be very difficult to read subscript qualifiers like 'Earth' and 'Ship'. We have the luxury of proper typesetting.

We have already met our starting point in special relativity – the light clock. As we discovered on page 121, this is just a beam of light travelling up and down between mirrors on a travelling spaceship, at right angles to the direction of travel. From the point of view of an astronaut on the ship, the light travels up and down vertically. But as far as a viewer on the Earth is concerned, the light beam travels diagonally at an angle, because the ship will have moved between the light leaving the top of the clock and arriving at the bottom.

All we need to produce the remarkable consequence of time dilation is a mental image of that clock, a spot of basic geometry and the idea that light goes at the same speed however you move.

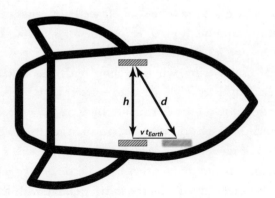

Figure 11: Light clock diagram.

We'll start by looking at the clock from the viewpoint of the astronauts on the ship. Let's say that the height of the clock is h. The speed of light is always referred to as c (because it's constant).

So the distance the light travels, h, will be t_{Ship} / c where t_{Ship} is the time that has elapsed on the ship.

Now let's jump back from the ship to Earth and watch that same event. We see the light beam taking a diagonal course. Thanks to Pythagoras' theorem we can work out the distance the light travels. Let's call it d. The other bit of our triangle is the distance the ship has moved forwards. That's the velocity of the ship, v, multiplied by the time that has elapsed from our earthly viewpoint, t_{Earth}. To keep our equations simple, we're going to use the convention that we don't need to put in a multiply sign, so v times t_{Earth} is just vt_{Earth}.

Then Pythagoras tells us that $d^2 = h^2 + (vt_{Earth})^2$. Given that light always travels at the same speed, we know that the distance d is just ct_{Earth} – the speed of light times the time it took. So that makes the equation:

$$(ct_{Earth})^2 = h^2 + (vt_{Earth})^2$$

We'll take away $(vt_{Earth})^2$ from both sides:

$$(ct_{Earth})^2 - (vt_{Earth})^2 = h^2$$

… which is the same as:

$$t^2_{Earth}\,(c^2 - v^2) = h^2$$

Now we divide both sides by the bit in brackets:

$$t^2_{Earth} = h^2 / (c^2 - v^2)$$

We're almost there. A little further up we said that h was t_{Ship} / c, so let's stick that in:

$$t^2_{\text{Earth}} = t^2_{\text{Ship}} / (c^2 / c^2 - v^2 / c^2)$$

which is

$$t^2_{\text{Earth}} = t^2_{\text{Ship}} / (1 - v^2 / c^2)$$

Finally, we take the square root:

$$t^2_{\text{Earth}} = t_{\text{Ship}} / (1 - v^2 / c^2)^{\frac{1}{2}}$$

... where that ½ means the square root of what's in the brackets. And that's it. We've just shown how time dilation emerges from the simple assumption that light travels at the same speed however you move. The formula above is the real thing, the time dilation formula from special relativity.

Now, admittedly, if you were never very comfortable with algebra, what we just did was a little messy and difficult to get the brain around. However, there is absolutely nothing there that a sixteen year old coping with maths in high school would not be comfortable with – and it took less than a page to come up with the result. This isn't, as it happens, how Einstein himself came up with the relationship. However, it is just as legitimate an approach and it demonstrates why it is bizarre that we don't cover relativity at school.

Now we can immediately see how travelling quickly enables a form of time travel by plugging a speed for the ship into the equation. Let's say the ship is going at a respectable half of the speed of light. So our time dilation equation gives us:

$$t_{\text{Earth}} = t_{\text{Ship}} / (1 - (c / 2)^2 / c^2)^{\frac{1}{2}}$$

… which is:

$$t_{\text{Earth}} = t_{\text{Ship}} / (1 - c^2 / 4c^2)^{\frac{1}{2}}$$

Let's cancel out the c^2:

$$t_{\text{Earth}} = t_{\text{Ship}} / (1 - \tfrac{1}{4})^{\frac{1}{2}} \quad \text{or} \quad t_{\text{Earth}} = t_{\text{Ship}} / (\tfrac{3}{4})^{\frac{1}{2}}$$

The square root of ¾ is around 0.866 so:

$$t_{\text{Earth}} = t_{\text{Ship}} / 0.866$$

… which is the same as:

$$t_{\text{Earth}} = 1.155 \, t_{\text{Ship}}$$

And there is our time travel. If the ship travels for ten years at this speed, when it gets back to Earth, 11.55 years will have gone by. The closer the ship gets to the speed of light, the closer that v^2 / c^2 gets to 1, so the number on the bottom of the equation gets smaller and smaller, making the multiplying factor bigger and bigger. The closer you get to the speed of light, the more time slows down on the ship as seen from Earth – the more the ship travels into the Earth's future.

It is possible to do something very similar with a light clock that is aligned with the direction of travel, rather than at right angles to it, to show how length is contracted for a moving object. The final prime aspect of special relativity, the increase in mass, is a little more fiddly, but perfectly possible by throwing in the conservation of momentum.

It ought to be stressed again that this light clock approach was not the one originally involved in deriving the effects of special relativity. Einstein and his contemporaries built on the impact of the Michelson–Morley experiment of 1887, which showed that two light beams at right angles in a moving frame of reference did not alter as the beams were rotated. The mathematics to explain this, and hence derive the basics of special relativity, is not a whole lot more complex, but it is not as intuitive as the light clock, which is why I prefer to use this example.

Notes

◇◇◇

Chapter 1

Page 6 – Bronowski's description of the intertwining of science and
culture is from Jacob Bronowski, *The Ascent of Man* (London: Book
Club Associates, 1976), p. 14.

Page 6 – Bronowski's comment that man is a shaper of the landscape is
from the opening of Jacob Bronowski, *The Ascent of Man* (London:
Book Club Associates, 1976), p. 19.

Chapter 2

Page 17 – The suggestion that a void can't exist because objects would
either stay at rest or keep moving is from Aristotle (trans. Edward
Hussey), *Physics Books III and IV* (London: Clarendon Press, 1983),
p. 35.

Page 17 – The original statement of Newton's first law is from Isaac
Newton (trans. I. Bernard Cohen and Anne Whitman), *The Principia:
Mathematical Principles of Natural Philosophy* (Berkeley and Los Angeles:
University of California Press, 1999), p. 416.

Page 18 – Roger Bacon's assertion that we couldn't see the stars if there
were a vacuum is from Roger Bacon (trans. Robert Belle Burke), *The
Opus Majus* (Philadelphia: University of Pennsylvania Press, 1928), p. 485.

Page 20 – Feynman's definition of physical laws comes from Richard
Feynman, *The Character of Physical Law* (London: Penguin, 1992), p. 13.

Page 21 – Feynman's comparison of physical laws to the rules in a game
of chequers comes from Richard Feynman, *The Character of Physical Law*
(London: Penguin, 1992), p. 36.

Page 22 – Steven Weinberg's description of a metaphorical ultimate book of physics is from an interview with the author, Brian Clegg, 'Rational Heroes: Steven Weinberg', *The Observer*, 3 March 2013.

Page 26 – Feynman's assertion that natural laws are simple is from Richard Feynman, *The Character of Physical Law* (London: Penguin, 1992), pp. 33–4.

Page 27 – Information on the measurement of the speed of light is from Brian Clegg, *Light Years* (London: Icon Books, 2015), pp. 136–43.

Page 29 – The story of the statistician whose friend doubts the use of pi is from Eugene Wigner, 'The unreasonable effectiveness of mathematics in the natural sciences', Richard Courant Lecture in mathematical sciences, delivered at New York University, 11 May 1959; *Communications on Pure and Applied Mathematics*, 13(1), pp. 1–14, 1960.

Page 31 – Information on observations suggesting a variation in the fine structure constant are from Michael Murphy, 'Are Nature's Laws Really Universal?' (2015, online), astronomy.swin.edu.au; available at: astronomy.swin.edu.au/~mmurphy/research/are-natures-laws-really-universal/ (accessed 30 October 2015).

Page 33 – Information on MOND (Modified Newtonian Dynamics) is from Brian Clegg, *Gravity* (New York: St Martin's Press, 2012), p. 229.

Page 33 – Failures to match observation for MOND and dark matter are discussed in Lisa Randall, *Dark Matter and the Dinosaurs* (London: The Bodley Head, 2016), pp. 17 and 305–6 respectively.

Page 35 – The two-dimensional 'romance' is Edwin Abbott, *Flatland* (New York: Dover, 1992), all pp.

Chapter 3

Page 48 – The 1398 reference to a particle in John Trevisa's translation of Bartholomaeus Anglicus' *De Proprietatibus Rerum* is quoted in the *Oxford English Dictionary* (2015, online); available at: www.oed.com/view/Entry/138255 (accessed 5 November 2015).

Page 50 – Feynman's emphasis that light behaves as a particle is from Richard Feynman, *QED: The Strange Theory of Light and Matter* (London: Penguin, 1990), p. 15.

Page 61 – Information on dark matter is from Lisa Randall, *Dark Matter and the Dinosaurs* (London: The Bodley Head, 2016), all pp.

Page 70 – The physicist Lisa Randall's description of 'ordinary matter chauvinists' is in Lisa Randall, *Dark Matter and the Dinosaurs* (London: The Bodley Head, 2016), p. 317.

Page 74 – Newton's definition of the new concept of 'mass' appears in Isaac Newton (trans. I. Bernard Cohen and Anne Whitman), *The Principia: Mathematical Principles of Natural Philosophy* (Berkeley and Los Angeles: University of California Press, 1999), pp. 403–4.

Page 75 – Details of the platinum–iridium kilogram standards are taken from Stephen Battersby, 'How to measure anything – and fix the foundations of science', *New Scientist* (2015 online); available at: www.newscientist.com/article/mg22830411-000-how-to-measure -anything-and-fix-the-foundations-of-science/ (accessed 30 October 2015).

Page 75 – The assertion that a kilo of French cheese is a microgram lighter than a kilo of Australian Vegemite comes from T. State, 'Can't take the heat? We need a universal measure on temperature', The Conversation (2015, online); available at: theconversation.com/cant -take-the-heat-we-need-a-universal-measure-on-temperature-47154 (accessed 23 October 2015).

Chapter 4

Page 84 – Information on and quote from St Augustine of Hippo are taken from St Augustine (trans., ed. and introd. by Henry Chadwick), *Confessions* (Oxford: Oxford University Press, 1998), all pp.

Page 84 – The identification of 'time' as the 55th most common word and most common noun in written English is from Oxforddictionaries. com, 'The OED: Facts about the language – Oxford Dictionaries' (2015, online); available at: www.oxforddictionaries.com/words/ the-oec-facts-about-the-language (accessed 14 September 2015).

Page 85 – Stephen Hawking's unfulfilled promise to answer the question of 'What is the nature of time?' is in Stephen Hawking, *A Brief History of Time* (New York: Bantam Books, 1998), p. 1.

Page 85 – The time traveller's description of the nature of time as a fourth dimension is from H.G. Wells, *The Time Machine* (London: Pan Books, 1965), p. 8.

Page 86 – The inconsistent reporting of the events during an apparent murder at university are described in Alex Boese, *Electrified Sheep* (London: Pan Books, 2012), pp. 145–8.

Page 88 – The earliest known use of 'Time is what keeps everything from happening at once' is taken from Project Gutenberg, 'The Girl in the Golden Atom' (2007, online); available at: www.gutenberg.org/files/21094/21094-h/21094-h.htm (accessed 14 September 2015).

Page 103 – Sean Carroll's metaphor of the arrow of time emerging from the proximity of the big bang as our concept of up and down emerges from the proximity of the Earth is from Sean Carroll, *From Eternity to Here* (Oxford: Oneworld Publications, 2010), pp. 30–2.

Page 104 – Carroll's assertion that we should seek timeless explanations is from Sean Carroll, *From Eternity to Here* (Oxford: Oneworld Publications, 2010), p. 43.

Page 105 – Lee Smolin's summary of his view of time is from Lee Smolin, *Time Reborn* (London: Allen Lane, 2013), p. xiv.

Page 106 – Smolin's assertion that taking a timeless view leads to making things up is in Lee Smolin, *Time Reborn* (London: Allen Lane, 2013), p. 11.

Page 106 – Einstein's quote about sitting on a stove and relativity comes in many variants (some have 'put your hand on a stove for two minutes', etc.); this is the original, the abstract of Albert Einstein, 'On the Effects of External Sensory Input on Time Dilation', *Journal of Exothermic Science and Technology*, 1(9), 1938.

Page 107 – Newton's assertion of the existence of absolute time is from Isaac Newton (trans. I. Bernard Cohen and Anne Whitman), *The Principia: Mathematical Principles of Natural Philosophy* (Berkeley and Los Angeles: University of California Press, 1999), p. 408.

Chapter 5

Page 112 – Feynman's assertion that we have no way to know how a rotating universe affects us is from Richard Feynman, *The Character of Physical Law* (London: Penguin, 1992), p. 97.

Page 132 – Einstein's paper deriving $E = mc^2$ is accessed at Einsteinpapers.press.princeton.edu; 'Volume 2: The Swiss Years: Writings, 1900–1909' (English translation supplement), p. 172 (2015, online); available at: einsteinpapers.press.princeton.edu/vol2-trans/186 (accessed 24 October 2015).

Chapter 6

Page 136 – The effects of lack of gravity on living organisms are described in Brian Clegg, *Gravity* (New York: St Martin's Press, 2012), pp. 6–7.

Page 137 – The suggestion that babies are experimenting with physics by dropping things is from Lynne Murray, *The Psychology of Babies* (London: Robinson, 2014), p. 218.

Page 138 – Galileo's dig at Aristotle for not experimenting is from Galileo Galilei (trans. Henry Crew and Alfonso de Salvio), *Dialogues Concerning Two New Sciences* (New York: Dover, 1954), pp. 62–3.

Page 139 – A video of Scott dropping a hammer and feather on the Moon to fall at the same rate can be seen at YouTube, 'Hammer vs Feather – Physics on the Moon' (2015, online); available at: www.youtube.com/watch?v=KDp1tiUsZw8 (accessed 16 October 2015).

Page 140 – Newton's apple story is recounted in the manuscript version of William Stukeley, *Memoirs of Sir Isaac Newton's Life* (London: Royal Society, 1752), p. 15ff; Royalsociety.org (2015, online); available at: royalsociety.org/collections/turning-pages (accessed 16 October 2015).

Page 143 – Newton's discussion with Robert Hooke on the hypothesis that an orbit combines a straight-line motion with falling is covered in a range of letters in Volume 2 (1676–1687) of Isaac Newton (ed. H.W. Turnbull), *The Correspondence of Isaac Newton* (Cambridge: Cambridge University Press, 2008).

Page 144 – Christopher Wren's offer of 40 shillings if Hooke could produce a proof of the inverse square law of gravitation is described in Andrew May, *Isaac Newton* (Stroud: The History Press, 2015), pp. 55–6.

Page 145 – The role of Ismael Boulliau in suggesting an inverse square law for the gravitational force is described in Thony Christie, 'The man who inverted and squared gravity', The Renaissance Mathematicus (2011, online); available at: thonyc.wordpress.com/2011/09/28/the-man-who-inverted-and-squared-gravity/ (accessed 21 November 2015).

Page 146 – Newton's 'moon test' for gravity is in Isaac Newton (trans. I. Bernard Cohen and Anne Whitman), *The Principia: Mathematical Principles of Natural Philosophy* (Berkeley and Los Angeles: University of California Press, 1999), pp. 803–5.

Page 149 – A discussion of the intention of 'hypotheses non fingo' is provided in the Guide to the *Principia* in Isaac Newton (trans. I. Bernard Cohen and Anne Whitman), *The Principia: Mathematical Principles of Natural Philosophy* (Berkeley and Los Angeles: University of California Press, 1999), pp. 274–7.

Page 151 – Huygens' dislike of the principle of attraction is noted in a letter from Huygens to Leibniz in Christiaan Huygens, *Oeuvres complètes, publiées par la Société hollandaise des sciences* (The Hague: Martinus Nijhoff, 1888–1950), Volume 9, p. 190.

Page 152 – The comments by Leibniz on attraction in gravity are from Gottfried Wilhelm Leibniz, *Die philosophischen Schriften* (Berlin: Weidmann, 1875–1890), Volume 5, p. 58.

Page 152 – Information on Newton's books is from John Harrison, *The Library of Isaac Newton* (Cambridge: Cambridge University Press, 2009), all pp.

Page 154 – Halley's introductory ode to the *Principia* is taken from Isaac Newton (trans. I. Bernard Cohen and Anne Whitman), *The Principia: Mathematical Principles of Natural Philosophy* (Berkeley and Los Angeles: University of California Press, 1999), p. 379.

Page 156 – John Gribbin's book on general relativity is John Gribbin, *Einstein's Masterwork: 1915 and the General Theory of Relativity* (London: Icon Books, 2015).

Page 157 – Einstein's description of having his 'happiest thought' while sitting in the Bern patent office is from his 1922 Kyoto lecture and referenced in W.F. Bynum and Roy Porter (eds), *The Oxford Dictionary of Scientific Quotations* (Oxford: Oxford University Press, 2005), p. 198.

Page 177 – Einstein's 1918 paper was 'On Gravitational Waves', accessed at: einsteinpapers.press.princeton.edu/vol7-trans/25 (accessed 12 February 2016).

Page 177 – Information on gravity waves is from Brian Clegg, *Gravity* (New York: St Martin's Press, 2012), pp. 223–49.

Chapter 7

Page 181 – The complaint about philosophy unweaving the rainbow is from John Keats, *Lamia* (1820); Bartleby.com (2015, online); available at: www.bartleby.com/126/37.html (accessed 21 November 2015).

Page 185 – The seven life processes are taken from Brian Clegg, *The Universe Inside You* (London: Icon Books, 2012), pp. 44–5.

Page 186 – Response to the question 'Are cells alive?' is from Dr Jennifer Rohn, originally quoted in Brian Clegg, *The Universe Inside You* (London: Icon Books, 2012), pp. 46–7.

Page 191 – Nick Lane's comments on the unlikeliness of ultraviolet as a source of energy for living things is from Nick Lane, *The Vital Question* (London: Profile Books, 2015), p. 93.

Page 193 – Nick Lane's assertion that it is pretty much impossible to distinguish a mushroom cell from a human one through a microscope is from Nick Lane, *The Vital Question* (London: Profile Books, 2015), p. 1.

Page 196 – The humorous book showing the discovery of a twentieth-century motel in the fortieth century is David Macaulay, *Motel of the Mysteries* (Boston: Houghton Mifflin, 1979), all pp.

Page 198 – Details of the use of zircons to date the earliest life are from Nick Lane, *The Vital Question* (London: Profile Books, 2015), pp. 24–6.

Page 198 – The 4.1-billion-year dating for the earliest known possible life is from Elizabeth Bell, Patrick Boehnke, Mark Harrison and Wendy Mao, 'Potentially Biogenic Carbon preserved in a 4.1 billion-year-old zircon', *Proceedings of the National Academy of Sciences of the United States of America*, PNAS 2015; published ahead of print, 19 October 2015: doi:10.1073/pnas.1517557112 (accessed 23 October 2015).

Page 202 – The quote on how 'damned similar eukaryotic cells are' is from Nick Lane, *The Vital Question* (London: Profile Books, 2015), p. 43.

Page 205 – Information on Fred Hoyle and panspermia is from Jane Gregory, *Fred Hoyle's Universe* (Oxford: Oxford University Press, 2005), pp. 286–8.

Page 210 – Nick Lane's examination of the origins of complex cells is covered in Nick Lane, *The Vital Question* (London: Profile Books, 2015), pp. 157–236.

Chapter 8

Page 218 – Edward de Bono described his picture of idea space and the tunnel in a workshop attended by the author in 1986.

Page 218 – Alex Osborn's techniques are often used without acknowledgement and are described in his book Alex Osborn, *Applied Imagination* (New York: Charles Scribner's Sons, 1979).

Page 219 – The use of creativity to distinguish humans from the (other) apes is from Mihaly Csikszentmihalyi, *Creativity: The Psychology and Discovery of Invention* (London: HarperCollins, 1996), p. 2.

Page 219 – Bohm's description of the loss of ability to apply different ways of looking at things is from David Bohm, *On Creativity* (Abingdon: Routledge, 1996), p. 4.

Page 220 – Bohm's description of the formation of a new basic order is from David Bohm, *On Creativity* (Abingdon: Routledge, 1996), p. 8.

Page 220 – The article describing chimps as more evolved than humans is 'Chimps Lead Evolutionary Race', *Nature* 446, 841 (2007).

Page 223 – The description of the creative process is from David Bohm, *On Creativity* (Abingdon: Routledge, 1996), p. 5.

Page 224 – The assertion that creativity involves 'going outside the framework of preconceptions' is from David Bohm, *On Creativity* (Abingdon: Routledge, 1996), p. 23.

Page 224 – Vera Rubin's description of suddenly seeing the data on star orbits in a galaxy differently is quoted in Mihaly Csikszentmihalyi, *Creativity: The Psychology and Discovery of Invention* (London: HarperCollins, 1996), p. 4.

Page 225 – Frank Offner's description of the benefits of not consciously thinking about a topic is from Mihaly Csikszentmihalyi, *Creativity: The Psychology and Discovery of Invention* (London: HarperCollins, 1996), p. 99.

Page 226 – Csikszentmihalyi's ideas on possible mechanisms for unconscious idea generation are from Mihaly Csikszentmihalyi, *Creativity: The Psychology and Discovery of Invention* (London: HarperCollins, 1996), pp. 101–2.

Page 226 – Csikszentmihalyi's description of the need for a change to be recognised and accepted before it is truly creative is from Mihaly Csikszentmihalyi, *Creativity: The Psychology and Discovery of Invention* (London: HarperCollins, 1996), pp. 27–31.

Page 227 – The story of Edwin Land and the development of the Polaroid camera is from Brian Clegg and Paul Birch, *Imagination Engineering* (London: FT Prentice Hall, 2000).

Page 227 – The suggestion that fun is the driver for creativity and there is evolutionary benefit to the race as a whole from this is from Mihaly Csikszentmihalyi, *Creativity: The Psychology and Discovery of Invention* (London: HarperCollins, 1996), pp. 107–10.

Page 234 – The Ishango bone is described in Amir Aczel, *Finding Zero* (New York: Palgrave, 2015), pp. 20–1.

Page 235 – Information on the development of counting is from Brian Clegg, *Are Numbers Real?* (New York: St Martin's Press, 2016), pp. 13–20.

Page 240 – The experiments by Fortunato Battaglia using transcranial magnetic stimulation on mice are described in 'Magnets Bolster Neural Connections', *New Scientist*, 26 May 2007.

Page 240 – The suggestion that new-grown neurons could be involved in memory formation comes from S. Ge et al., 'A Critical Period for Enhanced Synaptic Plasticity in Newly Generated Neurons of the Adult Brain', *Neuron* 54, 559–66 (2007).

Page 241 – More information on Theodore Berger's work on brain implant chips can be found at John Cohen, 'Brain Implants Could Help Alzheimer's and Others with Severe Memory Damage' *MIT Technology Review* (2013, online); available at: www.technologyreview.com/featuredstory/513681/memory-implants/ (accessed 5 January 2016).

Page 246 – The quote on the theory of Aristarchus, putting the Sun at the centre of the universe, is from Archimedes, *The Sand Reckoner*, in T.L. Heath, *The Works of Archimedes* (New York: Dover, 2002), pp. 221–2.

Page 254 – Henry Gee's criticism of Brian Cox's *Human Universe* is from Henry Gee, 'Brian Cox's *Human Universe* presents a fatally flawed view of evolution' (2014, Guardian Science Blogs); available at: www.theguardian.com/science/blog/2014/oct/14/brian-coxs-human-universe-presents-a-fatally-flawed-view-of-evolution (accessed 8 September 2015).

Chapter 9

Page 261 – The study of the impact of unequal societies on their citizens is Kate Pickett and Richard Wilkinson, *The Spirit Level* (London: Penguin Books, 2009), all pp.

Page 263 – Information on megastructures in the universe and attempts to explain them away is from Colin Stuart, 'Could cosmic megastructures be intruders from another world?', *New Scientist* (2015, online); available at: www.newscientist.com/article/mg22830440-300-could-cosmic-megastructures-be-intruders-from-another-world/ (accessed 4 November 2015).

Page 268 – The assertion that our universe seems strangely atypical is from Lee Smolin, 'You think there's a multiverse? Get real', *New Scientist* (2015, online); available at: www.newscientist.com/article/mg22530040.200-you-think-theres-a-multiverse-get-real/ (accessed 31 October 2015).

Page 276 – The relationship between mathematics and the natural world is explored in Brian Clegg, *Are Numbers Real?* (New York: St Martin's Press, 2016), all pp.

Further Reading

Applied Imagination, Alex Osborn (New York: Charles Scribner's Sons, 1979)

Are Numbers Real?, Brian Clegg (New York: St Martin's Press, 2016)

Creativity, Mihaly Csikszentmihalyi (London: HarperCollins, 1996)

Dark Matter and the Dinosaurs, Lisa Randall (London: The Bodley Head, 2016)

Dialogues Concerning Two New Sciences, Galileo Galilei, trans. Henry Crew and Alfonso de Salvio (New York: Dover, 1954)

Einstein: His Life and Universe, Walter Isaacson (London: Simon & Schuster, 2008)

Einstein's Greatest Mistake, David Bodanis (London: Little Brown, 2016)

Einstein's Masterwork, John Gribbin (London: Icon Books, 2015)

Flatland, Edwin Abbott (New York: Dover, 1992)

From Eternity to Here, Sean Carroll (London: Oneworld, 2011)

Gravity, Brian Clegg (New York: St Martin's Press, 2012)

How to Build a Time Machine, Brian Clegg (New York: St Martin's Press, 2011)

Isaac Newton, James Gleick (London: Harper Perennial, 2004)

Light Years, Brian Clegg (London: Icon Books, 2015)

On Creativity, David Bohm (Abingdon: Routledge, 1996)

Relativity: A Very Short Introduction, Russell Stannard (Oxford: OUP, 2008)

The Ascent of Man, Jakob Bronowski (London: Book Club Associates, 1976)

The Character of Physical Law, Richard Feynman (London: Penguin, 1992)

The Principia: Mathematical Principles of Natural Philosophy, Isaac Newton, trans. I. Bernard Cohen and Anne Whitman (Berkeley and Los Angeles: University of California Press, 1999)

The Spirit Level, Kate Pickett and Richard Wilkinson (London: Penguin Books, 2009)

The Strangest Man, Graham Farmelo (London: Faber & Faber, 2009)

The Vital Question, Nick Lane (London: Profile Books, 2015)

Time Reborn, Lee Smolin (London: Allen Lane, 2013)

Index

Index

Index

Index